Cyberschools:
An Education Renaissance
2010 Edition

By Glenn R. Jones

Please visit Cyberschools.com

INVITATION

Cyberschools function in a complex and incessantly changing ecosystem. Therefore, this book is meant to be merely the beginning of an extended, continuing discussion. All are welcome and encouraged to join the discussion at **Cyberschools.com**

Online you'll find more information about topics such as *Assessing Student Learning in the Knowledge Age: The Jones International University Assessment Model*, which was prepared by Dr. Joyce A. Scott and Dr. Robert W. Fulton, both associated with Jones International University at the time of its writing. Additional appendixes will be added to this book online as the discussion continues.

So that others can more freely use the contents of this book, and to support the open education resources movement, this edition of *Cyberschools* is licensed under a Creative Commons Attribution-Noncommercial-Share Alike 3.0 Unported License, except where otherwise noted.

First printing March 2010

ISBN 978-0-9826544-0-8

For information, please contact:
Jones International, Ltd.
9697 E. Mineral Ave.
Centennial, CO 80112

Cyberschools.com

Glenn R. Jones has spent four decades extending the reach of technology, first by bringing cable television to American homes, then by fusing education with the Internet to deliver education to lifelong learners worldwide. Along the way, Mr. Jones has created numerous businesses in the fields of digital encryption, digital compression, Internet technology, e-commerce, software development, education, cable networks, entertainment, mobile communications, radio networks and advertising sales. He has also authored several books.

Contents

Foreword   7

Acknowledgments   9

Special Comment   10

Glossary of Terms   11

Introduction: Cyberschools Enter the Mainstream ———— 13

> Two Wars Reshape America at the Dawn of the 21st Century 16
> Armed with Knowledge in the Trenches   18
> Technology as a Transforming Force   21
> Demand for Higher Education   22
> What the Knowledge Society Demands   23
> The Knowledge Worker   25
> The Brain Under Siege   26
> Critical Issues   27
> Asking Questions Beyond the Web   28
> The Lessons of Athens   29

Chapter 1: The Education Race ———————————— 31

> The New Adult Learner   35
> Cyberschools' Tipping Point?   38
> Global Learning Potential   42
> Older Students: Budget Minded Learners   45
> Who Higher Education Targets   46
> The Graying Demographic Trend   47
> Looking Forward   49

Chapter 2: The Costs/Benefits Equation ——————— 51

> U.S. Higher Education Meets the Bottom Line   52
> University Teacher Shortages   54
> Worker Retraining in a Knowledge Age: 55
  An International Market
> The Aging of Baby Boomers: Opportunity or Risk?   58
> A Nation Still at Risk? Education 25-Years Later   60

Chapter 3: The Roots of Cyberschools ——————— 65

> The Medium isn't the Message   66
> A Brief History   67
> Worldwide Distance Higher Education   69
> TV: Education Direct to Your Living Room   71
> Commercial TV's Early Forays into Education   72
> Public Television   73
> Watershed: The Public Broadcasting Act of 1967   74
> Community College TV   74
> Prognostication   74

Chapter 4: The Virtual Classroom of the 21st Century ——————— 75

> The Home as a Classroom   77
> Emerging Technologies in Higher Ed E-Learning   79
> Virtual Classrooms: Better than Real?   79
> Virtual Education Visions   82
> The Virtual Library   83
> The Decade "Google" Became a Verb   86
> The Transformation of Time Itself   88
> Teacher Holograms   89

Chapter 5: The Internet Comes to School ——————— 91

> The Web Whirlpool 93
> The Future is Here 95
> The Web 2.0 Explosion 97
> Worldwide Learning Revolution 98
> Livening up K–12 Education 100
> Online Learning in Canada 103
> The Explosion of Interest Internationally in Web 2.0 103
> The Question of Access 104
> International Connectivity Ramps Up 106
> What's Next? 108

Chapter 6: The Onrushing Revolution in Mobile Learning ———113

> Mobile Phones are "Lenses on the Online World" 115
> The M-Learning Boom 117
> Corporate M-Learning Case Study 118

Chapter 7: Distance Learning: Defining the Market ———121

> China's Education Market 123
> Europe's Demand and Challenges 125
> A Campus for Every Continent 128
> Worldwide Electronic Corporate Education and Training 129
> Who's Delivering Distance Corporate Education 130
> U.S. Universities Tap the Market 132
> The Global Course Prototype 133
> The Road Ahead 134

Chapter 8: Principles of Good Practice ———135

> Accreditation: Who Confers It? 138
> Worldwide Quality Standards 140
> What Accreditors Think 141

Chapter 9: Cyberschools and You ———————— 143

   ➚ A How-To Guide for Distance Leaners   143
   ➚ Distance Education, Not Instant Education   143
   ➚ A Rigorous Way to Learn   144
   ➚ Cyber Education, Not Passive Education   144
   ➚ A Better Way to Learn?   145
   ➚ Eight Questions to Ask   146
   ➚ Cyberstudent Opinions   148

Chapter 10: Free Market Fusion: One Path ———————— 153

   ➚ Defining Free Market Fusion   154
   ➚ Working Together   155
   ➚ Free Market Fusion, Entrepreneurs and Institutions   155
   ➚ Institutions: Inertia Versus Initiative   156
   ➚ Risk-Taking: The Key Role   159
   ➚ Combining Risk-Taking and Caution   159
   ➚ Modeling Free Market Fusion   160
   ➚ Structuring the Relationship   161
   ➚ Opportunities for Free Market Fusion   163
   ➚ Challenges are Plentiful   166
   ➚ The Larger Arena: Tapping Entrepreneurial Talent   166

Chapter 11: Epilogue ———————— 168

Appendix: Assessing Student Learning in the Knowledge Age   173

Footnotes ———————— 213

FOREWORD

TOWARD A LEARNING COMMUNITY ON THE PLANET

There are people who contribute greatly to society by thinking small—by focusing intensely on a set of narrow, sometimes incredibly complex, problems and working to solve them. There are others who contribute equally—and sometimes more—by thinking big. They tackle vast, recalcitrant, even more complex problems. Some, of course, do nothing but talk big dreams. Others work long and hard to make them come true and, for their trouble are often dismissed as cranks, dreamers, or idealists. Some are. A few, however, are just the reverse.

Glenn Jones is one of these. A successful, hardheaded businessman, he built a large, important cable TV, telecommunications, and new media company starting from absolute scratch. That would be dream enough for most people. But Jones believes that the most successful businesses of the future—bigger and more profitable than today's giants—will be those that help solve crucial social crises—environmental issues, health issues, and, above all, the educational crisis.

Although almost everyone is dissatisfied with existing schools, colleges, and universities—not simply in the U.S. but across much of the world—most proposed innovations take for granted that educational problems can be solved within the existing framework.

Are our factory-style elementary and secondary schools in trouble? Increase homework. Increase teachers' pay. Patch up broken windows. In short, make the factory run faster.

Even as new technology and a Third Wave, knowledge-based economy move us from mass production of goods to customized or individualized production, children are still subjected to mass production education. (The idea of replacing the factory-style school with reconceptualized alternatives is still regarded as heresy by the educational establishment.)

Higher education, too, requires deep reconceptualization. Dr. Donald Langenberg once speculated that "many universities may die or may change beyond recognition as a result of the IT [information technology] revolution.... Some may be 'virtual universities' that are delocalized across cyberspace." Which is what this book is about.

In public discourse, as in policy, "Education" with a capital E is regarded as a separate, specific category of social activity. "Media" are in another category. "Computers" are in still another category. Yet in the real world the boundaries among these categories are melting away. The world of computing and the world of media are converging. And it may be impossible to solve our most crucial social problems so long as we continue to think within the frame of these conventional categories.

Glenn Jones is a category buster. And we believe, as he does, that education cannot be brought into the Third Wave future so long as it is viewed as separate from both the media and cyberspace. In these pages, he lays out an exciting vision for the fusion of these activities into a worldwide education revolution. There is no single panacea for our problems. What Glenn Jones proposes cannot be expected to solve all the accumulated problems of our obsolete educational assembly lines. But it does offer high-powered tools to help people — not only the rich in the rich countries, but all people — gain access to better education at lower cost as a Third Wave learning community begins to form around the planet.

— Alvin and Heidi Toffler

Acknowledgements

The first edition of this book was published in 1997. Since then distance education has evolved dramatically. Increasingly, students choose to pursue academic achievement online. New studies and data illuminate the transformative effect of technology on education. But as far as we've come, we have only just begun.

In my quest to merge technology and education, I have met hundreds of leaders in business and academia, professors, teachers, UNESCO officers, high-technology practitioners, futurists, celebrities with a passion for education and innumerable policymakers and regulators around the globe. Without listing them individually, I sincerely acknowledge their contributions to the dream of making all the world a school.

Special Acknowledgement to Tom Billington

Cyberschools are inherently embedded in incessantly morphing technology and concepts. For this reason they have been prolific generators of discussion surrounding their potential, their economic model, their efficacy and their impact on traditional education models. Consequently, preparing this new edition required a large amount of time and research to bring the technology aspects current, to update statistics and to verify trends. Significant redrafting was required.

My partner in preparing this new edition of Cyberschools was Tom Billington. Tom not only did the research and much of the verification, but was extremely helpful in the resultant drafting and redrafting of content. He was also prolific in providing marketing concepts to distribute this book.

Tom did all of this with great enthusiasm and indefatigable energy. Without his help this new edition might never have reached the finish line. I am deeply grateful to him.

— *Glenn Jones*

From his early days developing the cable industry to his vision for worldwide distance learning, Glenn Jones has generated enormous changes in communications over the years. In this updated edition of *Cyberschools*, he shows us how far technology and online learning have traveled. He points to a future of limitless growth. We learn how virtual schools and universities can collapse the limitations of space, time, and distance.

As part of its historical perspective, *Cyberschools* references *A Nation at Risk* (ANAR), the 1983 report still recognized as a landmark education document. The report focused on high school education and the findings were wrenching to educators and shocking to business and civic leaders. There was a general mediocrity throughout the education system. Standards were low or nonexistent. Curriculum was an elective smorgasbord. Teaching was a low status profession. No one was being held accountable for poor student performance. Many students who graduated were unready for post-high school education.

In addition to these conclusions, ANAR delivered two significant messages. First, it emphasized that quality education remains the cornerstone of our stability, prosperity, and security. Second, our county does not have to put up with mediocre education. We have the tools and creativity to achieve excellence.

Cyberschools is an affirmation of these ideas. It provides a history of remarkable changes in technology and a powerful sense of present and future possibilities. It describes new modes of access to knowledge and information which cross boundaries of ocean and country. We learn about a "computer-mediated communication system" that addresses the challenges of equitable access and scale.

Cyberschools helps us see the centrality of technology in education reform at all levels. Its message is the transformation of worldwide interaction and learning. Glenn Jones shows us how a nation at risk can become a nation of promise and accomplishment.

— *Milt Goldberg*

In the dynamic, ever changing cyberschools sector, definitions abound. Hence, it is important at the outset to define some commonly used terms that will arise in this new edition of *Cyberschools*.

An educational "course" is defined in four ways by the Sloan Consortium (Sloan-C), a nonprofit consortium:

1. **A "Traditional" course**, using no online technology, has its content delivered orally and in writing.

2. **A "Web-facilitated" course** has one to 29 percent of its content delivered online. The course is essentially a face to face course that is facilitated using web based technology (such as a web page or course management system to post the syllabus).

3. **A "Blended/hybrid" course** has 30 to 79 percent of its content delivered online and blends online and face-to-face delivery. A substantial portion of the content is delivered online (typically using online discussions) and it usually has some face to face meetings.

4. Finally, and most importantly for this book, **an "online" course**, which has over 80 percent of its content delivered online, has no face to face meetings; and most or all of the content is delivered online.[1]

Cyberschools deliver "online" courses *only*, where 100 percent of the content is delivered via the Internet and no face to face meetings occur. They can include fully independent self-paced courses or semester-long, teacher-led courses. On a broader basis they can include fragments of the learning process created by any person or group, anywhere, for any purpose.

"Distance learning" or "distance education" refers to a form of education that can occur, but that doesn't *necessarily* occur, via the Internet. The U.S. Department of Education defines **distance education** as "a formal education process in which the student and instructor are not in the same place. Thus, instruction may be synchronous or asynchronous, and it may involve communication through the use of video, audio, or computer technologies, or by correspondence (which may include both written correspondence and the use of technology)."[2]

Electronic platform refers to the technology that makes the electronic delivery of education possible. It uses any of, or any combination of, a wide array of enabling devices and telecommunications systems, including broadcast, satellite, cable television, radio, landline or mobile telephones, the PC, digital devices and the Internet.

Clouds: *The Economist* stated that "much of computing will no longer be done on personal computers in homes and offices, but in the 'cloud': huge data centres housing vast storage systems and hundreds of thousands of servers, the powerful machines that dish up data over the Internet. Web-based email, social networking and online games are all examples of what are increasingly called cloud services, and are accessible through browsers, smart-phones or other 'client' devices."[3]

Asynchronous Learning is the "interaction between instructors and students [that] occurs intermittently with a time delay," according to the American Society for Training and Development (ASTD).

Synchronous Learning, on the other hand, involves "real-time, instructor-led online learning [events] in which all participants are logged on at the same time and communicate directly with each other."[4]

INTRODUCTION
Cyberschools Enter the Mainstream

*"There is only one good, knowledge,
and one evil, ignorance."*

Socrates, in *Diogenes Laertius*[5]

Much of my business life has been spent in the communications sector, and I have often used the following quote from renowned science fiction author, Dr. Isaac Asimov, in speeches I have given:

> "For a technological advance to be truly basic, it must change the entire world. In the last half century, the coming of television, of jet planes and of solid-state electronics has presented examples of world-changing technological advances. Dwarfing these, however, is the communications revolution."

I had the opportunity to speak with Dr. Asimov about these words on two occasions; once at the Library of Congress and once as a fellow guest on a friend's yacht. He held a broad view about the scope of the communications revolution.

I mention this because this book is mostly about fusing the communications revolution with education in order to create cyberschools. The discussion contained herein addresses the use of cyberschools in the more formal environment of the education process, i.e., K–12 and post-secondary education. This is done purposely to confine the focus. However, I wish to acknowledge that, one — there is already wide use of cyberschools in corporate training, which I mention later in the book, and two — that ad hoc cyberschools, created by anybody, anywhere, for any purpose will proliferate going forward.

Since the update of this book in 2002, much has happened. The relentless drumbeat of change continues at a quickened pace. The communications revolution has continued to fuse with education and we are now witnessing the confluence of mobile technology with the Internet. This is huge for education because mobile phones constitute, by far, the largest interconnected communications system on the planet. By the end of 2008, over four billion mobile subscribers existed worldwide.

Mobile phones such as the iPhone, the Google Android and the Palm Pre allow for particularly strong Internet access and other access enabling devices are bubbling into the market. This new edition of *Cyberschools* includes a new chapter entitled, "The Onrushing Revolution in Mobile Learning" to focus on this emerging, new area.

The amazing amount of multimedia offered on the Internet at such low cost, or for free, poses huge opportunities and risks. On one hand, it threatens the economic models of an array of media such as:

- Newspapers, where news or classifieds are now available for free on sites like Google News, Indeed.com and Craigslist;
- Book publishers, where self publishing has become easier than ever with the Internet;
- Book retailers, as Amazon.com cuts into in-store sales;
- Radio, where more and more stations, music and audio content of all sorts are accessible via the Internet; and

- TV stations, where the switch to IP will eventually redefine how they use their local bandwidth, how they monetize it, and how they keep it from being reallocated to others who want it.

On the other hand, the same industries have also seen upside from the Internet. Local newspapers can reach a global audience. So-called "backlist" books can continue to sell well online even though they are no longer in physical bookstores. Music can be sold individually on sites like iTunes, and new entrepreneurial businesses can begin with incredibly low startup costs generating revenue in a host of ways, such as with Google AdSense.

But make no mistake, mainstream media will never be the same. "This shattering of the mainstream into a zillion different cultural shards is something that upsets traditional and entertainment media to no end," wrote Chris Anderson in the influential book *The Long Tail*.[6]

Organizations that offer must-have products or services to distinct niches or that adapt their fundamental business models to the Internet are more likely to survive. To fail to adapt to the Internet is to face near certain death for many organizations, and even for many industries.

The educational system is not immune to these challenges. Certainly, since 2002 much progress has been made making classrooms Internet-accessible. While nothing replaces the power of a face-to-face inter-play between students and their teachers, the educational system generally has not kept up with the new generation, the so-called "Net Geners." As best-selling author Don Tapscott wrote in the book *Grown Up Digital*,

> "The Net Geners have grown up digital and they're living in the twenty-first century, but the education system in many places is lagging at least 100 years behind. The model of education that still prevails today was designed for the Industrial Age. It revolves around the teacher who delivers a one-size-fits-all, one-way lecture. The student,

working alone, is expected to absorb the content delivered by the teacher. This might have been good for the mass production economy, but it doesn't deliver for challenges of the digital economy, or for the Net Gen mind."[7]

The challenge is not just to keep up with the young but to continually adapt for the old, because the aging population both in the U.S. and countries like China is growing astronomically. The topics of lifelong learning and educating the new generation are addressed in this book.

Two Wars Reshape America at the Dawn of the 21st Century

Even as technology advances and continues to disrupt the education system, two massive wars continue to reshape America and the world. They constitute World Wars of a very different type than the two World Wars fought in the 20th Century.

The first war, of course, is the ongoing global war on terror that was precipitated by the attacks on September 11, 2001. The second is the global war on the economy that began seven years later. While many events led up to the global economic crisis, one date sticks out as the defining tipping point-date for this economic war: September 15, 2008, the date Lehman Brothers filed for Chapter 11.

The battle on terrorism and on the economy have some similar characteristics. The defining moments of each have their origins in New York City, the world's financial capital. They are marked by extreme uncertainty and fear. Sudden changes in either war can change the world literally, and with frightening consequences, within the blink of an eye; and they often result from perpetrators who are hard to identify and from problems that seem intractable.

September 2001 was, no doubt, an earthquake and shock to America's system that changed the country forever. The largest attack by terrorists on American soil altered the playing field dramatically. Later bombings in Madrid, London and Mumbai and the suicide bombings in Iraq and elsewhere continue to redraw our roles as educators, students, business people, entrepreneurs, professionals and citizens of the world.

Along with terrorism and the economy putting the world in such disarray, time itself seems to have been electronically modified. Responses are expected instantly through Blackberries, iPhones and other digital devices. Time is no longer a constraint via the Internet. Available 24/7, information is accessible globally and instantly; it is no longer available only at a certain time or place.

Technology has been both a blessing and a curse. The terrorists who bombed trains in Madrid in 2004 used their mobile phones to detonate the explosives, but a mobile phone then became the clue that uncovered the plot.[8] Technology has enabled a global network for terrorists to communicate, yet it has become critical in the search to bring them to justice.

Terrorists and those who support them would have us believe that theirs' is a unique system of beliefs that deserves unquestioned loyalty and acceptance by humankind. Destruction and suffering, oddly, seem to be the only means they can intellectually conceive and employ to advance their cause as their immediate tactic is not the improvement of humankind, but its reduction.

And their ultimate quest is not one of enlightenment and betterment for the world's billions of under-privileged and under-fed, but authoritarian power over them. The implied message is that no real education is required. Just do as we say and as we tell you to think and eternal happiness is assured. Deceptively simplistic, unfortunately the message only has to work with a very few to produce disastrous consequences.

While the perpetrators have waged war against innocents under the flimsy pretense of either religious or political righteousness (often both), the results have been something quite unforeseen for almost everyone concerned. Most importantly, those of us who are part of the new targets of violence have learned quickly to expect the unexpected.

"Unexpected" is also a good word to describe the economy. Since September 15, 2008 and into 2010, the economy continues to shift and change with wild uncertainty. Large, once highly valued companies such as AIG, Fannie Mae, Chrysler and General Motors, have witnessed unprecedented government intervention.

These two wars have also demonstrated the consequences of globalization. What happens in the U.S. has immediate repercussions at major centers worldwide and vice versa. It's an always-on and increasingly connected world. The Internet with its ability to transmit information instantly is a major catalyst for this globalization, a trend that has profound implications for the educational system as will be discussed later in this book.

ARMED WITH KNOWLEDGE IN THE TRENCHES

We have also come to understand that while an armed response may be an essential first reaction and critical on-going defense in the war on terrorism, the long-term solution to the war will be to arm individuals with knowledge and the intellectual tools of modern education. The soft power of education, rather than armed power, will ultimately have far greater influence in the wars. The ethics and importance of planetary understanding outreaches every religious belief and political pretense.

As beneficiaries of the most well-endowed civilization in history, our greatest challenge and opportunity will be to provide teaching environments and educational tools that will enable economic improvement and a better quality of life across the spectrum of global society. As Thomas Jefferson said,

> "Enlighten the people generally, and tyranny and oppressions of body and mind will vanish like evil spirits at the dawn of day."[9]

We already have the means and abilities to answer this challenge, and now we must demonstrate the will to persevere. We must make sure that technology ultimately serves the forward progress of knowledge and humankind. These are the ultimate defenses against and eventual victors over terrorism and war. How we go about delivering them are crucial questions that require our immediate attention.

Since Socrates first addressed the nature of knowledge, educators have struggled with two central questions: what to teach and how to teach it. The debate echoes from antiquity and volleys across today's mainstream bestsellers, exploding beyond the borders in town halls, legislatures, corporate boardrooms, and even dining rooms throughout the world.

Included in the idea behind cyberschools in this book is the delivery of education to people instead of people to education. This concept provides tested and proven answers to the third and fourth questions: where should the learning take place and what constitutes learning?

In our current global society, how these questions are answered has special significance because the answers will have a major impact on the many youthful, vibrant experiments with self-government, including the delicate ongoing experiment with self-government in the U.S. As historians Will and Ariel Durant pointed out more than two decades ago in *Lessons of History*, access to education is the key:

> "If equality of educational opportunity can be established, democracy will be real and justified. For this is the vital truth beneath its catchwords: that though men cannot be equal, their access to educational opportunity can be made more nearly equal." [10]

Our global culture is undergoing a transformation that lends great urgency to these questions. We are imbedded in a headlong race through an information technology revolution to a knowledge-based society.

And in this headlong race, certain industries will fall short or break. Many are struggling to create workable Internet models. In 2008, the newspaper industry, for instance, suffered some serious blows. One example, the Tribune Company, the nation's second largest newspaper publisher responsible for the respected *Chicago Tribune* and *Los Angeles Times*, declared Chapter 11 bankruptcy in December of 2008. The print media has ceded much to digital media—with profound implications for education.

The last time *Cyberschools* was updated in 2002, the huge growth in Internet users and PC usage were showcased. Indeed, as is shown in Table 1, that growth continues unabated.

Table 1. Growth in Personal Computers and Internet Users [11]

New forms of social collaboration and social networking worldwide

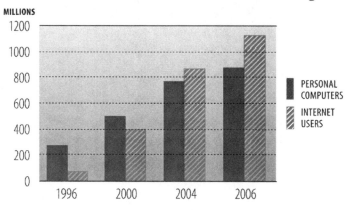

are growing as the Internet and wireless devices continue to grow in usage. Blogs, wikis, social networking sites such as Facebook and MySpace are similarly exploding in usage. And, of course, search itself has universalized the access to knowledge spurred by the "granddaddy" of all search engines, Google, and its growing number of competitors such as Bing and Yahoo!

Mobile devices can be the largest distribution vehicle for the Internet, and search is the most important research tool. But search is much broader than that really. As journalist John Battelle stated,

> "Search is no longer a stand-alone application, a useful but impersonal tool for finding something on a new medium called the World Wide Web. Increasingly, search is our mechanism for how we understand ourselves, our world and our place within it. It's how we navigate the one infinite resource that drives human culture: knowledge."[12]

This new technological universe and explosion of knowledge-enabling information have also created a sense that we are out of control. Decision makers who must in any event make decisions are deluged with information they cannot grasp and choices they do not comprehend. Everything is moving with great speed. We are inundated with information yet starved for knowledge. There seems to be nothing to hold onto, and yet we must marvel at what is occurring.

This type of chaos echoes in the opening lines of the poem, "The Second Coming," by William Butler Yeats:

> "Turning and turning in the widening gyre
> The falcon cannot hear the falconer;
> Things fall apart; the centre cannot hold;
> Mere anarchy is loosed upon the earth."[13]

TECHNOLOGY AS A TRANSFORMING FORCE

In a process that is unique in the history of humankind, we have created technological capabilities that are rapidly transforming the globe. Landline telephones in the early 1900s and television transmissions in the 1950s provided some rough context, but those technologies were quickly placed under the lock and key of traditional power interests. Perhaps the most important aspect of today's technologies is they proliferate beyond the control of traditional regulatory and industry cliques, though no small effort and expense are being expended to change that.

Most amazing, these changes have defied experts and pundits alike by progressing outside the bounds of any grand design. They have been galvanized in many cases by the availability of a staggering array of electronic devices ranging from Blackberries to PDAs to pagers to iPhones. The change has come about at the grassroots level of the world's societies, implemented by people using these new tools to communicate about how and why to accomplish the things they care about. And as previously indicated, these changes threaten whole industries.

Again using the newspaper industry as an example, witness how they have moved toward social media and digital reading devices. In 2008, more than 30 newspapers around the world provided content to Amazon.com's Kindle e-reading device for a monthly subscription fee.[14] And, one signal of the ascendance of social media is that Donald Graham, the Chairman and C.E.O. of The Washington Post Co., known for its pioneering print journalism, joined the Board of Directors of Facebook in 2008.

DEMAND FOR HIGHER EDUCATION

Because of the rapid rise in literacy and education levels in many countries, North American, Western European, and Australian higher education has become increasingly attractive to foreign students. In 2006-2007, for instance, there were about 583,000 foreign students studying at U.S. colleges and universities. Fifty nine percent of those students were from Asian countries.[15]

Students are willing to travel great distances and in many cases commit themselves to years of government service in exchange for the financial support to enroll in traditional U.S. Canadian, European, and Australian degree programs on campus. We must think about how the opportunities for these students, and millions more who do not have the financial means for overseas study, will exponentially expand if they can begin to receive the same quality and content of coursework via electronic means, without the travel and at a lower cost.

Conversely, this electronic capability will make it possible for the great Asian universities to reach out to the rest of the world — a critical factor to improve the world's understanding of Asian culture.

The demand for electronic delivery of all kinds of courses at all levels, including corporate training and K–12 classes, will mushroom in the 21st century. Our global society's rate of technological adaptation is both driving that demand and providing the tools to present education via electronic media.

WHAT THE KNOWLEDGE SOCIETY DEMANDS

To understand this chicken-and-egg phenomenon, it is important to recognize some of the fundamental characteristics of our new knowledge society and the unique attributes that some Western-educated workers lend to it.

The knowledge society we have entered differs greatly from the industrial society we leave behind. In the industrial society the principal resource was energy, and its tools were artifacts like forklifts, cranes, trucks, trains, automobiles, and airplanes. Its principal characteristic was that it allowed us to extend the human body.

The knowledge society is different because the velocity of its evolution is much more rapid, and its principal resource is information. As has often been noted, information is a special kind of resource. It can be weightless, invisible, and in many different places at once. The tools of the knowledge society drive the creation, storage, delivery, manipulation, and transformation of that information. Importantly, the principal characteristic of the knowledge society is that it allows us to dramatically extend the human mind by, among other things, introducing a new model for learning. This acceleration of change has increased dramatically such that 10 years of change today could be the equivalent of 100 or more years of change in the past.

The quantum extension of the human mind combined with the ability to extend the human body has resulted in a new reality. A reality in which the human mind, excluding religion and acts of nature, is now more clearly the most powerful force on the planet.

We are all integrally involved in this evolution, and education has an important part to play. Education is a process. Education is how information becomes meaningful. Information without meaning is useless. Education converts information into knowledge, understanding, and wisdom much like changing temperature turns water into ice. Education is the loom through which information is woven into value systems, dignity, self-worth, freedom, and into civilization itself. It can help to eradicate the ignorance and hatred surrounding the 9/11 crisis, lead to much greater knowledge that can help solve the economic issues of the future, and evolve civilization in a positive way.

Agile, quick and ever evolving as a species, humans have been constantly interacting and changing in relation to their environments and to new tools. Man began as *Homo habilis* or the "handy man" surviving with primitive tools such as stone or bone, then evolved into *Homo erectus* or the "upright man" taming fire for its use, and evolved further into *Homo sapiens* or the "wise man" who has had breakthroughs such as the wheel and language. The next evolution, as a special issue of the *Economist* states, may be *Homo mobilis* or the man on the move, untethered and yet in constant touch with his fellow man through technologies that bind man together, versus splitting them apart. [16]

There may well come a time when one's brain could be augmented with neural chip implants. To illustrate, Google Co-Founder Larry Page reportedly said, "on the more exciting front, you can imagine your brain being augmented by Google. For example you think about something and your cell phone could whisper the answer into your ear." [17]

THE KNOWLEDGE WORKER

All societies contribute to the evolution of education. However, economic and workforce experts claim that the developed world's universities produce graduates with unique capabilities. The emergence of the "knowledge worker" college graduate was first defined by Peter F. Drucker in his 1959 book, *Landmarks of Tomorrow.*[18] The later reexamination of the term by Robert B. Reich in his 1991 book, *The Work of Nations: Preparing Ourselves for 21st Century Capitalism*, goes a long way toward describing the reasons why demand for receiving college curriculum course content from the developed world's universities is so high. The knowledge worker, or symbolic analyst as labeled by Reich, is that person who can produce new designs and concepts, as opposed to following standard procedures and producing familiar products.

Reich describes the demand for such an education:

> "Millions of people across the globe are trying to learn symbolic-analytic skills, and many are succeeding. Researchers and engineers in East Asia and Western Europe are gathering valuable insights into microelectronics, macrobiotics, and new materials, and translating these insights into new products. Young people in many developing nations are swarming into universities to learn the symbolic and analytic secrets of design engineering, computer engineering, marketing, and management."[19]

These continue to be sought-after educational experiences. Perhaps someday a society on a distant planet will seek to import the educational offerings that make possible our world's symbolic analyst learners.

The evolution and continual specialization of our education processes in no way lessens the overall importance of education for the world as a whole. In fact, they intensify it. Will and Ariel Durant stated it well:

"Our finest contemporary achievement is our unprecedented expenditure of wealth and toil in the provision of higher education for all." [20]

THE BRAIN UNDER SIEGE

The importance of general education is growing in tandem with the world's body of knowledge. Our libraries strain under the weight of books sounding the alarm about an information revolution and the speed at which new information is being generated. The holdings of the world's libraries are increasing dramatically, and much of the content is being digitized. The amount of information conveyed electronically is higher than at any time in world history.

Between the end of 2005 and the end of 2007, for example, the number of annualized text (SMS) messages alone grew from 81 billion to 363 billion, according to CTIA-The Wireless Association®. In fiscal year 2007, the Library of Congress had 614 million page views of its web site, and the Library's online historical collections included 13.6 million digital files. [21] The fusion between education, information and technology is clearly happening at quantum speed.

At the end of the day, this information must be dealt with by an electrochemical contraption that weighs three pounds, more or less, takes up about half a cubic foot of space, runs on glucose at about 25 watts, processes information at the rate of approximately 100 quadrillion operations per second, looks like a big walnut, and is the world's first wet computer: the human brain. That brain is under siege, bombarded from all sides by torrents of new information and from every direction through continually evolving, efficient digital devices.

The brain has become electronically dependant as it seeks to absorb the massive flood of information made available by a dizzying array of new digital devices, not to mention billions of web sites, blogs, wikis, and other Internet-enabled media. The cyberschool approaches described in this book embrace an array of tools with which this vast body of information can be considered, managed, and put to use by individuals. These tools can empower individuals by giving them the means to convert information into knowledge, understanding, and wisdom.

They are technologies that can help spread out the decision-making process in governments, institutions, and businesses. It is obvious that technological advances have created a communications environment in which vast amounts of information can be delivered inexpensively, an environment in which the barrier of distance is erased and the barrier of time is diminished.

We are blazing through a transition that is technology-driven. If we don't measure up to the responsibilities of leadership in the concepts and content we select and develop to distribute through our new technology, the knowledge society will fall far below its potential.

Revolutions are transitory. Technologies will continue to be invented and evolve, but the fundamental restructuring of the world's economic and political systems already has been set in motion. The result is rapidly unfolding around us, a knowledge-based society that is the legacy ascendant of the information revolution, proliferated by the communications revolution.

CRITICAL ISSUES

Critical issues must be addressed to ensure the quality and speed of our educational processes and content. This reality was first stated in economic terms by William B. Johnston and Arnold H. Packer in their landmark study *Workforce 2000*, which cited education and training as the primary systems by which human capital is both developed and protected. The speed and efficiency with which these systems transmit knowledge and influence the rate of growth in human capital are more important than the traditional gauge of rate of investment in plant and equipment, the same study noted. [22]

The sequel book called *Workforce 2020* had three broad conclusions when it was published in 1997. First, the labor market of 2020 will demand highly educated workers who can work with advanced technologies while low skilled work can be done anywhere in the world. Second, the intense competition and globalization will create "intense volatility" in work. And, finally, the American workforce will become slightly "more brown and black in the next twenty years, but its pervasive new tint will be gray."[23]

The change and volatility is worldwide. It is the obligation and opportunity of every person and organization committed to the concept of self-government and to the forward progress of civilization to lend what tools they can to assist in the education of humankind. The challenges are what, how, where, and when to teach for optimum benefit.

ASKING QUESTIONS BEYOND THE WEB

The cyberschool is a dynamic learning environment that requires the full engagement of the faculty and students who use it. Such a school is much more than web pages and streams of emails. It must be supported by a technical and administrative cadre who interact with faculty and students constantly in order to ensure that a quality education is the final outcome.

Online education and traditional on-campus education are blending together seamlessly as a result of technology enhancements. Web facilitated and blended/hybrid courses are very common. Many universities have also created totally online, or cyberschool, programs. It is the cyberschool aspect of traditional schools and new for-profit schools that are disruptive in many ways — including their ability to deliver education to people who could never access it before, in ways they have never seen before, and with a power and influence in their lives that they never dreamed of.

As a step in this direction, I would like to suggest two questions for students and educators to ask themselves and their colleagues as they pursue electronic education approaches and evolve their own technological capabilities at the dawn of the new millennium:

- Can we learn to embrace change while continuing to respect our traditions?
- Are we prepared to deal with shifts in societal demands and the new technologies that will dwarf the changes we have witnessed to date?

THE LESSONS OF ATHENS

The truly disruptive nature of the web allows forward-thinking entities to thrive while wiping out others unwilling to adapt. Chris Anderson, Editor-in-Chief of *Wired* magazine, described how a company like Google has so successfully adapted itself to the changing nature of the web. Citing author Paul Graham, he writes:

> "The Web naturally has a certain grain, and Google
> is aligned with it. That's why their success seems so
> effortless. They're sailing with the wind, instead of sitting
> becalmed praying for a business model, like the print
> media, or trying to tack upwind by suing their customers,
> like Microsoft and the record labels. Google doesn't try
> to force things to happen their way. They try to figure out
> what's going to happen, and arrange to be standing there
> when it does." [24]

In the knowledge age, besides taking stock of our strengths and playing to them, educators have to be incredibly nimble in adjusting to change when, not if, it occurs. The founder of Taoism Lao Tzu wisely said, "If you do not change direction, you may end up where you are heading." [25]

Imagine the vibrant energy and intellect of Athens during the time of Socrates and Plato. The ghost of Athens is visible today. It has been said that Plato, in all his strivings to imagine an ideal training school, failed to notice that Athens itself was a greater school than even he could dream of.

Let us notice our environment. It is time now to continue fusing our electronic tools with all manner of teaching entities and information repositories. It is time to create a world that is, like Athens was, a great school, a world vibrant with interest and excitement about education, a world where educational opportunity is visible to all and where hope is alive. A world that sees the wilderness of information and technology as its new frontier.

To do this, we need cyberschools.

CHAPTER ONE
The Education Race

"There is no domestic knowledge and no international knowledge. There is only knowledge."

Peter F. Drucker in The Atlantic Monthly [26]

Since the last update of Cyberschools in 2002, globalization has continued its historic advance. And this global stampede has put greater pressures on educational institutions, particularly in the U.S., to keep up. Most important, it has spotlighted the need for a highly adaptable and educated workforce. Of course, this wonderful need to be continually adaptable is nothing new for the U.S. As the 12th Librarian of Congress Daniel Boorstin wrote back in 1978,

> "Nothing is more distinctive, nor has made us more un-European than our disbelief in the ancient well-documented impossibilities. Every day we receive invitations to try something new. And we still give the traditional, exuberant American answer: 'Why not!'." [27]

Educational enrollments worldwide are truly impressive. In 2005, about 1.3 billion students were enrolled in schools worldwide. Of that number, 693 million were in elementary-level programs, 511 million in secondary programs and 138 million in higher education. Between 1994 and 2005, in Africa alone, enrollments in elementary education increased 68 percent while they increased 97 percent and 191 percent in secondary and higher education respectively. [28]

U.S. universities consistently rank among the highest in the world. In fact, in 2008, 13 of the top 20—and three of the top 5—universities in the world were U.S based, according to the Times Higher Education-QS World University rankings. [29] Demonstrating the rise of Asian institutions, nine of the top 50 universities were Asian-based; and two new entrants to the top 50 list were both Asian: the Hong Kong University of Science and Technology and Seoul National University. [30]

The dawning of any century can be counted on to elicit prophecies and prognostications from all manner of philosophers, poets, and pontificators. For the world of higher education, the predictions are that schools and universities worldwide are going to be faced with educating more people with fewer dollars for longer periods of time and then finding them back again a few years later for more.

History provides the evidence that these forecasts are right. In essence, higher education in the 21st century is finding itself playing a game of catch-up, and the U.S. is falling behind. The destruction and death wrought by World War II in Europe, Asia, and Africa left those continents and their societies years behind in their development of key competencies, including educational infrastructure and programs.

Beginning in the late 1960s—reflecting the 20-plus years it took for many countries just to produce new college-age generations—the world's educational institutions began to respond to new demands. The following three decades saw the demand for education evolve at an incredible rate. Between 1980 and 2005, the number of students worldwide from preschool through all types of higher education increased by 485 million or 57 percent from 857 million to 1.342 billion.[31] The greatest percentage growth occurred in the college, university, technical and vocational training enrollments. Total worldwide enrollments reported increased from 51 million in 1980 to about 138 million in 2005.[32]

In the U.S. in the Fall of 2008, over 74 million people were enrolled in American schools or colleges, according to estimates from the U.S. Department of Education.[33] Enrollment in public elementary and secondary schools rose 26 percent between 1985 and 2008.[34] There also has been a tremendous and encouraging increase in the enrollments of primary and secondary students, particularly in developing countries.

Yet the increase in numbers only tells part of the story. While the U.S. expends more dollars on education than any country in the world, it falls considerably behind other countries in some fundamental ways, particularly in math and science. The OECD surveyed 15-year olds in the principal industrialized countries through its Programme for International Student Assessment (PISA). In 2006, 15 year old students in the U.S. scored *lower* on science literacy than their peers in 16 of 29 OECD countries.[35]

There is some good news that the world's literacy level is improving steadily. In fact, between 1990 and 2004, the number of illiterates declined by nearly 100 million people, mainly due to the decline of nearly 94 million in China alone.[36] The global illiteracy rate, however, is still very high. Between 2000 and 2004, 771 million adults were still illiterate, representing 18 percent of the world's total adult population. And efforts to reach the illiterate are increasingly hampered by the lack of computers. In 2007, Mexico had a student-to-computer ratio of one to 21; Guatemala had one to 71; Malawi and Niger had one to 3,000; and less than 10 percent of schools in many African and Latin American countries have Internet access, according to

Newsweek.[37] Where Internet education isn't available, radio and the rapidly evolving cell phone (with its many advanced applications) can serve the place of a PC.

In this age when the number of illiterates is so high, Internet learning provides huge opportunities. "Once disparaged as the jurisdiction of 'diploma mills' and profiteers," stated *Newsweek*, "the Internet is reforming this image: there's an explosion of new Web-based teaching tools made available to struggling school systems, from free open-source curriculums to online networks for refugee children trying to keep up with their classwork."[38]

Table 2. Illiteracy and Literacy Snapshot[39]

Year	1990	2004
World Illiteracy (000)	871,750	771,129
World Literacy Rates	75.4 percent	81.9 percent

While illiteracy remains high, enrollments continue to increase throughout most of the world; the thirst for education remains robust. Between 1990 and 2005, elementary enrollment increased 68 percent in Africa, 15 percent in Asia and eight percent in North America. Enrollments at the secondary level increased by 164 percent in Central America and South America and 97 percent in Africa.[40]

Most astounding is that the number of students seeking higher educa-tion—including vocational training and university certificate and degree programs—more than doubled during the same 15-year period, adding more than 87 million students. That's the highest growth rate of any single educational sector.[41]

Driven by world population growth, improving literacy rates, and desires for growth in personal incomes, the demand for higher educa-tion continues to grow *(Table 4).*

Table 3. World Education Vital Statistics [42]

	1995	2005	Percent Increase
Population:	5.69 billion	6.47 billion	14 percent
Students *(Elem.-Higher)*	1.02 billion	1.34 billion	31 percent
Students *(Higher)*	81.6 million	138 million	69 percent

THE NEW ADULT LEARNER

Secondary school graduates are by no means the only segment of the world's population seeking higher education. Let's look at how the world's student body has changed over the past generation.

Our past assumptions about who the typical college student was and how, when, why, and where that student attended college are no longer valid. Today the world's colleges and universities are faced with new student body demographics. This trend coincides with the arrival of the digital age.

In the U.S. more women attend colleges than men; many older, non-traditional students are earning degrees while raising families; and by 2020, estimates are that students of color will comprise 46 percent of the total student population. [43] And we no longer define a college education as something we do between the ages of 18 and 22. We have come to understand and embrace the concept of "lifelong learning." [44] Indeed, lifelong learning has moved from the category of "discretionary" personal investment to "essential" as people scramble to bolster their credentials in a volatile global workplace. So lifelong learning has advanced from 'nice phrase' to a business performance "imperative," as the book *Workforce Crisis* stated citing words from GE's Jack Welch in his final annual report to shareholders in 2000:

> "The most significant change in GE has been its
> transformation into a Learning Company. Our true 'core

competency' today is not manufacturing or services, but the global recruiting and nurturing of the world's best people and the cultivation in them of an insatiable desire to learn, to stretch and to do things better every day." [45]

Lifelong learning is now recognized globally as a key to a company's and a country's competitiveness. The European Commission, for instance, stated that,

"It is time to guarantee the place of adult learning in lifelong learning and to secure its role at all levels of policy making, so that its contribution to meeting Europe's challenges can be realised." [46]

Upward mobility through education is not just a tactic of white-collar management and computer professionals either. Professionals looking for educational options range from nurse practitioners to golf course groundskeepers and from sanitation and environmental technicians to assistant chefs.

For a number of reasons (one being an aging Western population and another being the view outside the industrialized world that college education is an import, versus export, industry), many countries now are turning their attention to providing education that will keep their college students and their educational dollars at home. Why ship them out to Los Angeles or Paris, they ask?

Another interesting dimension to the growing demand for education is that although the developing world's student population is young, the priorities placed on education and continued improvement by all societies suggest that this large and rapidly growing young student body represents a long-term market for lifelong learning programs. Some educational authorities describe the "bubble" of Western baby boomers who are demanding lifelong learning options as though it is a phenomenon that might disappear into retirement homes in the next 30-years. This is a mistake.

In some academic circles, there resides the faulty assumption that, within a generation or so, the world's student bodies will have returned to traditional colleges because their campuses are the only bona fide educational source. This assumption fails to take into account both the characteristics of the rest of the world's lifelong learners and the societal, technological and economic forces that will drive them over the next 20-years. At the very least, lifelong learners among the Western baby boomers will soon be joined and eventually eclipsed by seekers of education from developing countries in both East and West Asia and, eventually, Africa.

To meet both the short-term and long-term demand, countries must either build universities and staff them with world-class faculty or augment their higher educational institutions with less expensive alternatives. Cyberschools are a powerful alternative.

And they have become an ever growing and attractive form of learning. Convenient, accessible, cost effective for the provider and the student, and scalable, cyberschools are continuing to grow. As this type of learning further improves in quality, the teacher shortage exacerbates, the cost differential between site-based schools and cyberschools become more visible; and as costs fall faster, the outlook for cyber-schools seems bright indeed.

Table 4. World Higher Education: Student Growth (thousands of students) [47]

	1995	2005	Percentage Change
Africa	3,966	8,312	110 percent
Asia	30,890	62,600	103 percent
Europe	21,047	31,787	51 percent
Central/South America	8,455	15,575	84 percent
Northern America	16,026	18,613	16 percent
World	81,552	138,183	69 percent

CYBERSCHOOLS' TIPPING POINT?

The tipping point for cyberschools may be here. Since the publication of the last edition of *Cyberschools* in 2002, distance education has grown dramatically.

From 45,000 enrollments in fully online or blended-online courses in the fall of 2000, that number grew nearly 22 times to one million by the fall of 2007. While that amount represents only one percent of all courses in 2007, Harvard Business School Professor Clayton Christensen in his provocative book, *Disrupting Class,* predicts by 2019 about 50 percent of high school courses will be delivered online. [48]

Consider these other statistics:

- 42 states have significant supplemental, or full-time, online learning programs or both.
- 26 state-wide or state-led virtual schools exist in the U.S, according to the latest statistics available at press time.
- 173 virtual charter schools were serving 92,235 students in 18 states as of January 2007.
- 57 percent of public secondary schools provide access to students for online learning.

In 2006, an estimated 3.2 million post-secondary students in the U.S. took at least one online course, representing a 25 percent increase over the prior year. [49]

Within distance education, cyberschools are a relatively new concept because they offer educational content and classes that are conducted electronically. They are definitely an economical option.

Despite the growth of online education, much money has been lost, and many failed ventures have occurred to get to this point. Much of the early hype, in fact, for e-learning was unrealistic, and many executives and administrators made unwise decisions. At no time was the hype greater, or the crash harder, than during the dot com crash and the bubble that burst.

In the university sector Columbia University's FATHOM and New York University's NYU Online crashed during 2001–2002. Many ventures, kept alive by venture capital cash or simply hopeful dreams, did not survive.

While the venture capital market lost untold dollars during the dot-bomb and many technology start-ups were eviscerated, out of the ashes emerged several transformative companies such as Amazon.com, eBay and perhaps the most startling of them all: Google. Similarly, e-learning rose from the ashes. As Patti Shank from Learning Peaks, LLC, an instructional design and technology consulting firm, stated in *The e-Learning Handbook*:

> "When most technologies are introduced, they pass through a phase of inflated expectations, or hype. When expectations prove to be unfounded, an inevitable disillusionment follows. At the depths of disappointment, the real possibilities for the technology emerge... Having moved past the hype, we can now really consider using e-learning in transformative ways." [50]

Cyberschools have appeared at an opportune moment in history. The global education budget in 2004 in U.S. dollars stood at $1.97 trillion. [52] Some education leaders insisted that education at all levels—and accredited courses in particular—be delivered only by providing traditional bricks-and-mortar campuses with class ratios and class environments approved by faculty committees and administrators. These educational Luddites completely missed the boat. They fail to understand the global societal changes driving educational demand and what must be done to respond to it.

The quality of educational content and delivery has been and should always be the first concern. Likewise, there always will be a place for traditional campuses and classroom settings. But 21st century students need varied classroom environments and diverse educational delivery systems. There is no one way courses must be taught, so long as students learn and can demonstrate their learning through accredited testing and examination procedures. And the revolution in new communication devices has led to untold opportunities for the education sector.

The growing competition for dollars in education in developed and developing countries alike makes it crucial that new models for education incorporate delivery by both traditional institutions and carefully integrated electronic platforms.

Table 5. 2006 Higher Education Students (millions) [53]

	Population	Higher Ed. Students	Percent in Higher Ed.
China	1,314	23.36	1.8 percent
India	1,112	12.85	1.1 percent
Indonesia	232	3.65	1.6 percent
Japan	128	4.09	3.2 percent
South Korea	48	3.21	6.7 percent
Thailand	654	2.34	3.6 percent
U.S.	298	17.49	5.9 percent

Although the level of public financing for higher education has plateaued in many countries and in some countries has even decreased concurrently with unprecedented demand for education, alarmed pundits ignore three crucial facts:

- First, students of enormously varying financial means are finding ways to attain higher levels of education, often without the public financing support of a decade ago.

- Second, in this era of public budget deficits, public financing for education — in most countries — is constrained and certainly is not likely to increase at the same rate as the demand for education, no matter how shrill the alarms. The private sector, from small private universities and polytechnics to electronic colleges offering online courses and degrees, is already offering alternatives.

- Third, in many cases, especially with developing countries, public funds are being redirected to pre-primary, primary, and secondary education and away from higher education in an effort to improve basic literacy and student income-producing capability.

There is good reason to argue that public funding cannot, and need not keep pace, even at the secondary level. The U.S. in particular spent immense amounts of money on education during the 1970s, 1980s and 1990s, with dubious results.

The worker shortage, however, is real. The book *Workforce Crisis* states that, "the U.S. will need 18 million new college degree holders by 2012 to cover job growth and replace retirees, but, at current graduate rates, will be six million short." [54]

But is more public financing of education the answer to alleviating the crisis? The facts generally don't support this conclusion. Cyberschools can play a critically important role in alleviating the crisis, as this book will continue to detail.

Of the nearly two trillion dollars spent on education worldwide, the U.S. is currently the world's single greatest investor in education. Its public education budget is close to that of all the governments in six regions combined: the Arab States, Central and Eastern Europe, Central Asia, Latin America and the Caribbean, South and West Asia and sub-Saharan Africa. [55]

To put it another way, the U.S. which is home to just 4 percent of the world's children and young people, spends 28 percent of the global education budget. [56] Despite this massive outlay of funds to support the educational system, huge problems persist, and U.S. students often perform at or below the level of those in other advanced economies, as previously explained.

But test scores only tell part of the problem. A report produced for the Bill and Melinda Gates Foundation in 2006, *The Silent Epidemic*, stated one of the most stark issues: dropouts.

"There is a high school dropout epidemic in America. Each year, almost one third of all public high school students—and nearly one half of all blacks, Hispanics and Native Americans—fail to graduate from public high school with their class. Many of these students abandon school with less than two years to complete their high school education. This tragic cycle has not substantially improved during the past few decades when education reform has been high on the public agenda. During this time, the public has been almost entirely unaware of the severity of the dropout problem due to inaccurate data. The consequences remain tragic." [57]

Clearly, the government has not been getting the job done and isn't the sole answer. Some degree of competition needs to be introduced to the system; the private sector needs to introduce new models and methods of learning; and changes need to occur. Yet the government is most often resistant to change while the private sector almost thrives on change, being adaptable, nimble and flexible.

Cyberschools can provide a needed answer. Consider the cost to go to a Harvard graduate school versus the cost of getting a degree online. Then consider the added cost of taking the necessary time off work to complete a degree at a university when a person can get the same degree online, at a lesser cost, from the convenience of one's home, and with excellent results.

GLOBAL LEARNING POTENTIAL

Distance education holds great potential to provide education among many third world countries, even though many obstacles remain. [58]

The International Association for K–12 Online Learning (iNACOL), under the leadership of Susan Patrick, conducted a survey in 2006 of e-learning in the K–12 market internationally, and its portrait of two countries at very different stages of development is fascinating. In China, online education still accounts for a very small percentage of China's population, but its growth is fast. According to the report,

> "Due to high costs and the current state of the Internet,
> e-Learning in China is still new and only used as a
> supplement to the current face-to-face content. Currently,
> there are 67 universities participating in a pilot to run
> online programs in both large and mid-sized cities.
> Online education as a new and modern educational form
> is growing fast. In 1999, the online educational market
> amounted to $1.7 billion U.S. dollars, while in 2004 it
> rose to $23.1 billion U.S. dollars, and the average growth
> rate was 66 percent, a very high increase." [59]

In Singapore, on the other hand, e-learning is very advanced. "As of November 2006, all (100 percent) of secondary schools and junior colleges and 134 (85 percent) primary schools (grades 1–6) are using an LMS [Learning Management System] for teaching and learning," according to iNACOL. "A number of schools in Singapore have adopted e-Learning week, where students do not attend school but stay at home working on lessons and assignments delivered through the learning management system. During this week, teachers facilitate the learning and provide feedback via email and other electronic means… Students are using wikis, blogs, and photo blogs to aid in the reflections of their learning, as well as incorporating student profiles into the LMS to build community," according to the report. [60]

The lack of Internet access is, however, a large impediment to much of the developing world's ability to engage in e-learning. UNESCO issues an annual report on "Education for All" (EFA) which addresses the aim of universal education globally. That report in 2008 reported that, even though over one billion people now have access to the Web, "the Internet remains inaccessible to most children, youth and adults in the countries that are struggling the most to achieve EFA." [61]

India shows how distance education can help fill the gap that traditional schools can't fill in developing countries. In 2004 India launched EDUSAT, the world's first dedicated education satellite, devoted exclusively to beaming distance learning courses, according to UNESCO. EDUSAT is a collaborative project of the Indian Space Research Organisation, the Ministry of Human Resources, state departments of education and the Indira Gandhi National Open University. A UNESCO report stated that,

> "Its aim is to improve and expand virtual learning for children, youth and adults by providing connectivity to schools, colleges, higher levels of education and non-formal education centres. A year after its launch, virtual classrooms had become a reality, with the connection of more than a dozen teacher training centres and fifty government schools in Kerala state." [62]

Ruwan Salgado, Executive Director of World Links, a not-for-profit enterprise dedicated to providing disadvantaged youth around the world with access to information technology, added that, "Developing countries, such as India, have increased expenditures six-fold and so are beginning to make significant financial commitments and investments in teaching information technology skills in their schools. The implication is that, in the next decade, developing countries, such as India, will make significant progress in narrowing the gap between the U.S. and themselves in providing their students with information technology skills, both in breadth and quality." [63]

Even if e-learning overall is proving slow in terms of uptake in many countries, institutions clearly feel they should be offering it. Almost all institutions studied by the OECD have some form of central strategy for e-learning or were in the process of developing one. In one survey in 2004, only 9 percent of 122 Commonwealth institutions lacked an institution-wide online learning strategy or plans to develop one, down from 18 percent in 2002. [64]

OLDER STUDENTS: BUDGET MINDED LEARNERS

For at least some of the U.S. higher education establishment, economic reality has arrived in the form of a degreed student who must retrain to keep a job and doesn't have time or money for campus frills. When the first edition of this book was prepared in mid-1996, researchers found more than one source that predicted enrollment of the "traditional" student in colleges and universities would dwindle. As late as 1979 in the U.S., traditional full-time students, 18 to 22-years old and usually straight out of high school, numbered 4.5 million. One source predicted that by 1992, enrollment of traditional students would fall from the 1979 high down to 3.1 million, a decline of 32 percent. [65]

Statistics available in mid-2000 showed that not only did the decline not occur but there had been a considerable increase in demand for education for both full-time and part-time students. By 2006, 7.6 million students under 22-years old were enrolled in degree-granting institutions. That number is predicted to rise to about 8.1 million by 2017. [66]

The most encouraging news is that more adults than ever before are seeking an education, motivated by an increasingly technology-driven employment marketplace that has less and less room for those who try to compete with just high school level skills. Also, the demand for part-time college-level education has grown right along.

While the number of young students has grown more quickly than the number of older students in past years, this trend will likely begin to shift dramatically. From 1995 to 2006, the enrollment of students under age 25 increased by 33 percent while the students over age 25 increased only 13 percent. From 2005 to 2017, the National Center for Education Statistics (NCES) predicts a rise of 19 percent in enrollments of people 25 and over. [67]

Adult learners typically have little interest in the expensive "extras" of college such as social and athletic events, association with sororities or fraternities, and various other on-campus organizations and activities.

They need flexible scheduling, affordable prices, and attendance options. In many cases, the existence of college libraries and bookstores are conveniences they will gladly forgo, providing they can receive reference materials and study assignments by postal service or, increasingly, over the Internet. Such students exist in all types and levels of education, and they are found virtually all around the globe.

Ken Dychtwald and Joe Flower in their book *Age Wave* described the changing approach to education in the late 1990's.

> "You may stop working one or more times in your thirties, forties or fifties in order to go back to school, raise a second (or third) family, enter a new business, or simply to take a couple of years to travel and enjoy yourself. You may go back to work in your sixties, seventies, or even eighties. You may find that the traditional framework of life — with youth the time for learning, adulthood for nonstop working and raising a family, and old age for retirement — will come unglued, offering new options at every stage. A cyclic life arrangement will replace the current linear life plan as people change direction and take up new challenges many times in their lives." [68]

WHO HIGHER EDUCATION TARGETS

Traditionally, most North American universities target graduating high school students as their prime market for recruiting. With lower birthrates since 1970, that pool of higher education prospects has stabilized. The economic crisis beginning in 2008 has forced many people to put off their retirements as they have seen their retirement accounts dwindle in value. At the same time, another dynamic has been at work. Those who earned their degrees or began earning them in the 1960s, '70s, '80s and '90s are now changing careers, seeking more training and education to sustain and excel in their chosen fields, and making plans to work beyond the once mandatory retirement age of 65. To do this, they are looking for accessible degree completion programs and new career education opportunities.

And North American employers are contributing to this trend by extending retirement ages in order to fill critical labor force shortages and keep on tap a highly experienced pool of talent. The advertising industry is beginning to respond to the graying of the world's population as well. The target age span demographic for many Madison Avenue agency brand campaigns has now been broadened from the 18 to 49 range all the way to 54. If Mick Jagger, a 66-year old grandfather, can still pack a stadium, who can afford to ignore the trend?

THE GRAYING DEMOGRAPHIC TREND

We know that the graying demographic trend is not limited to North America and Europe. Political economist and demographer Nicholas Eberstadt described China this way:

> "Its elderly population will be exploding in the years
> ahead... Between 2005 and 2025, about two-thirds of
> China's aggregate population growth will occur in the 65+
> grouping—and that cohort will likely double in size, to
> roughly 200 million people. By 2025, under current UN
> and Census Bureau projections, China would account for
> less than a fifth of the world's population but almost a
> fourth of the world's senior citizens." [69]

On the other hand, while India's 65+ cohort is expected to double in size between 2005 and 2025, those elders will account for only eight percent of the overall population 20 years after. This will leave India a relatively youthful country. [70] Yet the aging trend overall must be underscored. Adds Eberstadt:

> "The current and impending graying of Asia and Eurasia is
> an all but inevitable force, since it is propelled by the basic

arithmetic of longer lives and smaller families, trends, we
will recall, that have already been developing in the region
for decades if not generations. Only a catastrophe of
biblical proportions could forestall the tendency for Asia's
population to age substantially between now and 2025."[71]

At the same time, falling or stabilizing birthrates in East Asia mean
there will be fewer young students graduating from institutions to
nudge their elders into retirement and help support them once they
do retire. One of the solutions will be to continue to educate and
retrain older adults to keep them in the Asian workplace. This would
require a conventional educational infrastructure several times the
size of what the region now supports in order to keep older adult
productivity at a desirable level. The revolution in communications
can turbo-charge this growth. Cyberschools delivered via the Internet,
cell phones, and e-readers can help supply some of this infrastructure
quickly and at a comparatively low cost as an alternative to bricks-
and-mortar classroom buildings.

Although West Asia, Africa, and Central Asia currently have different
population dynamics, they can be expected to eventually follow similar
trends as they find the right formulas to match educational investments with human capital needs and follow the path to prosperity.
In North America, one interesting trend has remained consistent
as the student age demographic has changed. Older adult students
still want the same things they did when the postal service was the
only means of distance education delivery: education access at an
affordable cost.

Germany and Japan present the most obvious examples of varying
attitudes toward education—varying compared with the norms of
the U.S. and the U.K.—that have been successful in terms of standards of living and GNP. Both countries surpass the U.S. and British
standards of living, yet each has different approaches to education
and workforce training. Based on that country's long tradition of

apprenticeship, in Germany, workers expect and receive considerable ongoing training throughout their careers, most paid for by the government and their employers.

In Japan, the country's rigorous secondary school system produces graduates with what some estimate is the equivalent of a U.S. four year college education. This explains why large Japanese companies expect new employees, just out of high school, to be capable of immediately completing an engineering course of study before assuming their places on the factory floors. [72]

Other European and Asian countries have developed similar systems, and most are now continuing to integrate continuing education into their workforce training programs. As industrialized nations transform into knowledge-based economies and developing countries undertake mainline manufacturing, higher education institutions worldwide must make comparable shifts in the way they deliver their educational products. It is a difficult transition that must simultaneously address technology adaptation and confront deeply entrenched perceptions on more traditional campuses.

LOOKING FORWARD

Many trends signal that e-learning's steep growth will continue. Countries, companies and huge numbers of individuals are profoundly aware of the globalization of the workforce and the increasing worldwide demand for education, particularly in countries like India and China. The economic impact of where the best trained workers reside is increasingly obvious. The relationship between education and economics is visible for all to see.

The ability and economic advantage of reaching large bodies of people electronically through the Internet and wireless devices is becoming increasingly important. The advantages of cyberschools in offering a lower cost, high convenience, often privately-financed, high-quality and interactive education are more and more undeniable. They embrace, not resist, evolving technology.

The world is in an education race. Whoever wins the education race wins the economic race. This is true from a country's standpoint and from a company's standpoint. It is just as true from an individual's viewpoint.

In this environment, cyberschools are crucial.

CHAPTER TWO
The Costs/Benefits Equation

The price tag for a four-year undergraduate degree and related expenses now can run higher than $200,000.[73]

It is not an overstatement to say that a college education is becoming what it was 100 years ago: prohibitively expensive to all but the world's most well off and the minor number who receive significant financial aid. Yet the importance of receiving a strong higher education could not be more important. As Will and Ariel Durant stated it,

> "If education is the transmission of civilization, we are
> unquestionably progressing. Civilization is not inherited;
> it has to be learned and earned by each generation anew."[74]

U.S HIGHER EDUCATION MEETS THE BOTTOM LINE

Higher education is a powerful part of the U.S. economy. By 2007, 18.2 million students were enrolled, and about 3.6 million people were employed, in colleges and universities. [75]

However, U.S. higher education is an increasingly troubled $373 billion industry. [76] And those troubles were exacerbated greatly by the economic crisis that erupted in 2008. While other countries' higher education systems also face severe budget constraints, I offer this focused look at the U.S. education dilemma because it is a bellwether for the world's other education markets and, if solutions emerge, can offer a paradigm for change.

Perhaps the most pressing concern regarding higher education is the astounding increase in the cost of attending college. The 1980s saw health-care costs rise a whopping 117 percent, but the average cost of attending a public college increased by nearly 200 percent. [77]

The rise continued from 1995 to 2005. Over that period, after adjusting for inflation, tuition and fees rose on average 36 percent at private four-year colleges and universities, 51 percent at public institutions and 30 percent at community colleges. [78] As college costs continue to outdistance inflation, their rapid increase effectively denies educational opportunity to those unable to afford the escalating expense.

Look at the plight of low income families. While the total cost of attending a four year public institution represents 39 percent of a low income family's earnings, the cost of attending a private institution represents 69 percent of their annual income. Yes, 69 percent! Average annual tuition and fees for attending a four-year private university rose from $19,000 in 1987–1988 to over $32,000 in 2007–2008. [79] Some institutions with stronger endowments can offer substantial scholarships and financial aid, yet many cannot. The statistics represent a national crisis. Education is one of the few industries in the U.S.

that has become less rather than more productive. This is not a minor issue, because higher education is the engine for continued creativity, entrepreneurship and innovation in the coming decades.

Enrollments also continue to escalate. While enrollment in the 10 years from 1985–1995 grew 16 percent, it grew at the faster rate of 23 percent in the following decade.[80] States, furthermore, are facing budget cutbacks and will need to cut funds at higher education institutions.

As mentioned, colleges and universities are facing multiple crises stemming from the economic downturn of 2008. Consider these examples cited in the *New York Times* realizing, of course, that endowment shortfalls and employment levels can swing back with a resurging stock market and economy: in early 2009, the University of Florida eliminated 430 faculty and staff positions; Arizona State University eliminated over 500 jobs, announced it would close 48 programs and every employee would have between 10–15 days unpaid furlough days.[81]

Some other spending by higher education is simply frivolous. Wrote a reporter for *Money Magazine*:

> "If colleges were spending most of their money on
> initiatives that improve the quality of education for
> students, you might regard price hikes running at two
> to four times the rate of inflation as a necessary evil. But
> spending on palatial dorms, state-of-the-art fitness centers
> and a panoply of gourmet dining options? Maybe not."[82]

At the same time access to higher education for geographically distant students — those who must travel, and others who cannot attend campus-held classes — has become a high priority. As Americans recognize the importance of a college education to their careers, their quality of life, their economy, and their children's futures, they are increasingly concerned about universal access to higher education.

In an increasingly competitive world economy, the U.S. cannot let people with potential drop out of the educational system. If, indeed, they do, they also drop out of the economic system at ever-more unpalatable costs to society.

The need for universal higher education was made by Thomas L. Friedman in his award-winning book *The World is Flat*:

> "Everyone should have a chance to be educated beyond high school. Otherwise upper-income kids will get those skills and their slice, and the lower-income kids will never get a chance. We have to increase the government subsidies that make it possible for more and more kids to attend community colleges and more and more low-skilled workers to get retrained. JFK wanted to put a man on the moon. My vision is to put every American man or woman on a campus." [83]

To take this one step further, my dream is to put every man or woman on the planet into a virtual campus for at least some of their educational experience.

UNIVERSITY TEACHER SHORTAGES

Another change affecting higher education is the teacher shortage. The aging teacher population in higher education is one cause of the teacher shortage. In 1999, for instance, 29 percent of teachers were over 50 years of age; by 2007, that figure shot up to 42 percent. This suggests, as Clayton Christensen points out in his book *Disrupting Class*, that a decade hence there will be a wave of teacher shortages across the country even as the student population, the highest it's ever been, will not likely decline. [84]

As the children of the baby boomers move through U.S. colleges and universities, they will expand demand for faculty at the same time that many professors, hired to meet the baby boom demand of the

1960s and 1970s, are scheduled to retire. Unless a means is found to deliver education to more students without radically increasing the number of faculty, many would-be students will be closed out of the higher education system.

WORKER RETRAINING IN A KNOWLEDGE AGE: AN INTERNATIONAL MARKET

Business and labor leaders recognize the importance of retraining workers with skills that meet 21st century employment needs. And online learning is an increasingly acceptable, some say even better, way to become retrained. John Zogby, the famous pollster, wrote a provocative book *The Way We'll Be: The Zogby Report on the Transformation of the American Dream*. In one poll he conducted, 43 percent of 1,545 CEOs and small business owners nationwide agreed that a "degree earned through an online or distance-learning program is as credible as a degree earned through a traditional campus-based program."[85]

Online education improves upon campus learning in several ways. It requires no travel, can be conducted at a fraction of the cost and has all sorts of quality checks.

The educational imperative for workers applies to both industrialized and developing countries and their workers. Those with the highest education, not surprisingly, generally make the most money. Across 25 OECD countries and Israel, individuals with university degrees and advanced research education had earnings that were at least 50 percent higher than students whose highest educational level was below the upper secondary level. [86]

In America, the tie between earnings potential and education is equally strong. Between 1975 and 2003, real average earnings for adults aged 25 to 64 increased by a greater percentage at each higher level of educational attainment achieved past high school, according to the National Center on Education and the Economy. So while high school graduates saw a one percent decrease in real average earnings in that period, college graduates saw an average 19 percent increase, and graduate or professional degrees saw a 46 percent increase. [87]

Furthermore, unemployment rates are markedly higher for the less well educated. In 2004, the unemployment rates stood at 8.8 percent for those with less than a high school diploma, at 5.5 percent for those with a high school degree or GED degree, and fell under three percent for those with a master's degree. [88]

For 15 of the 30 fastest growing occupations, a bachelor's or higher degree is the most significant source of postsecondary education or training. [89] There is a consensus that nearly 50 percent of U.S. workers are employed in some aspect of the "knowledge," or "information," economy, although the definition of what constitutes "information work" is undergoing continued re-evaluation. [90]

Consequently, the U.S.' competitive edge in what is now a global marketplace is based on its ability to teach workers not just to be technically proficient, but to think, to evaluate, to adapt, to use information resources, and to become lifelong learners. A similar transformation is under way in the European Union countries, Canada, Japan, and Singapore. These skills are critical in all areas of industry, not just among top-level management. As Alvin and Heidi Toffler wrote in the foreword to the last edition of *Cyberschools*,

> "Even as new technology and a Third Wave, knowledge-based economy move us from mass production of goods to customized or individualized production, children are still subjected to mass production education. (The idea of replacing the factory-style school with reconceptualized alternatives is still regarded as heresy by the educational establishment.) Higher education, too, requires deep reconceptualization." [91]

The *Economist* noted the importance of education to the knowledge economy:

> "The world is in the grips of a 'soft revolution' in which knowledge is replacing physical resources as the main driver of economic growth.... The best companies are now devoting at least a third of their investment to knowledge-intensive intangibles such as R&D, licensing and marketing. Universities are among the most important engines of the knowledge economy. Not only do they produce the brain workers who man it, they also provide much of its backbone, from laboratories to libraries to computer networks." [92]

The challenges to anyone competing in today's global economy are great; and the turbulence is massive. Economists estimate that as many as 40 million people were dislocated by the "restructuring" in world manufacturing from 1980 to 1995. [93] In the 1990s, another 3.1 million people were laid off by corporate America, according to the Chicago-based outplacement firm Challenger, Gray, and Christmas. More restructuring will take place as these companies respond to marketplace demand and further competitive threats.

The economic crisis beginning in 2008 has created the greatest turbulence in the economy—and in the educational institutions that support it—since the Great Depression. The collapse of the housing market involving sub-prime mortgages and derivatives led to a massive banking crisis that reached its tipping point on September 15, 2008. The once legendary Lehman Brothers filed for Chapter 11 bankruptcy, and the brokerage firm Merrill Lynch was subject to a $50 billion takeover by Bank of America.

At that point, fear and uncertainty rippled through the national and global markets with tornado-like force. That day, the Dow Jones closed 504 points down, or 4.4 percent, the largest decline since the re-opening of the stock markets after 9/11. Major media reflected the fear the day after. "Fears of a global financial meltdown grew yesterday as the world's biggest bankruptcy plunged markets into turmoil," began an article in The Times Online on September 16, 2008. [94]

The fallout was like a forest fire rapidly spreading out of control, and the impact on the global economy was scorching. The government continued to step in with the full array of tools at its disposal. As the Dow lost over a trillion dollars in value, over a trillion more was dedicated to rescuing the global economy from utter collapse. Failing banks, insurance companies like AIG, and once hallowed companies like Fannie Mae and Freddie Mac posted unprecedented losses. Layoffs escalated; the Dow Jones tumbled; consumer confidence cratered; and liquidity and the banks' willingness to lend reached near Great Depression lows.

One can only guess how much of an impact the tremendous deficits and economic turmoil will have. One thing's for sure though: if September 11, 2001, kicked off the U.S. war on terrorism, September 2008 kicked off the country's even more costly new battle: a global economic battle fought in the economic trenches.

THE AGING OF BABY BOOMERS: OPPORTUNITY OR RISK?

Besides the economic changes, other demographic issues are occurring. The workforce, as noted, is increasingly aging; and companies are slow to realize the potential of the older workers. The tectonic shift in demographics in the U.S. workforce was chronicled by *The Harvard Business Review* (HBR):

> "In the next decade or so, when baby boomers — the
> 76 million people born between 1946 and 1964, more
> than one-quarter of all Americans — start hitting their
> sixties and contemplating retirement, there won't nearly
> be enough young people entering the workforce to
> compensate for the exodus." [95]

HBR profiled a sample company called RWE Power, Europe's third largest energy utility, as a microcosm of the issues facing many companies and organizations nationwide. "Today, some 20 percent of the division's workforce is over the age of 50. Projections indicate that this age group will make up more than half the workforce by 2011 — and close to 80 percent by 2018." [96] In the U.S. energy sector, over a third of the workforce is already over 50-years, and that age group will likely grow by more than 25 percent by 2020. [97]

Baby boomers will be a force in the economy for years to come. Called by HBR "the most financially powerful generation of mature consumers ever," today's mature adults control more than $7 trillion in wealth in the U.S. — 70 percent of the total. Between 2000–2010, the growth in the U.S. workforce is expected to fall by 10 percent in the age group of 35-44 while the growth between 55–64 will increase by 52 percent. [98]

While training can play a critical role, it is often focused at younger, rather than older, employees. Older workers (age 55 plus) reportedly receive on average less than half the amount of training that any of their younger cohorts receive, including workers in the 45 to 54 range. [99] For anyone engaged in continuing professional development or online learning, this poses a huge opportunity, of course, to provide more training for older workers and the baby boomers. And those over 55 need to take charge of their own training.

Training can play a key role in the retention, recruitment and development of talented, older employees. Yet while many corporations excel in corporate training, many do not effectively engage in training for employees to make whole-sale changes in jobs at an older age. Instead, they offer a whole series of euphemisms for ways to lower costs by getting older, more expensive employees to retire: buyouts, restructurings, rightsizing, upsizing, etc.

And the nation's economy is moving increasingly to a service economy requiring often higher level skills while the more commoditized skills can be outsourced outside the U.S. at far less costs. From 2006 to 2016, service-providing industries will generate almost all of the employment gain and will provide more than three-quarters of all jobs in 2016. [100] Clearly, new tools and concepts are required to master this rampant change in our environment. Higher levels of education are imperative for workers in every upward-developing country.

The turbulence of change and the need to adjust are manifest. Yet, for the worker who needs retraining, the military man or woman, the rural adult learner, shift workers, homebound parents, and the gifted high school student with no opportunity to take college-level classes at his or her high school, access to educational opportunities generally—and to college coursework and credit specifically—has been difficult, if not impossible.

Indeed, as I have noted, at the end of the 20th century, higher education was close to becoming the privilege it was at the century's beginning. This does not need to be the case, however. Higher education for the smallest number would be a tragic legacy for us to leave succeeding generations required to compete in the global economy of the 21st century.

A NATION STILL AT RISK? EDUCATION 25-YEARS LATER...

More than twenty five years have passed since the publication of *A Nation at Risk*, a publication during the Reagan era by a blue ribbon commission.

A Nation at Risk remains today perhaps the most often referenced education report in the nation's history. "It called for higher standards for all students, improved teacher status and quality, and responsible leadership for elected officials at all levels of government," said Milton Goldberg, former Executive Director of the National Commission on Excellence in Education which produced *A Nation at Risk*. "The report emphasized that education was key to the quality of life in our country. This belief undergirds present and recent efforts to make a first-rate education accessible to all," Dr. Goldberg added.

On the 25th Anniversary of *A Nation at Risk*, Chester Finn in *The Wall Street Journal* wrote that since its publication, some progress has been made. "Instead of judging schools by their services, resources or fairness, we track their progress against preset academic standards — and hold them to account for those results. We're also far more open to charter schools, vouchers, virtual schools, home schooling."[101]

Yet despite these changes, Chester Finn argues, "our school results haven't appreciably improved, whether one looks at test scores or graduation rates. Sure, there are up and down blips in the data, but no big and lasting changes in performance, even though we're also spending tons more money."[102]

Elementary and secondary schools in the U.S. have had to grapple with shifting circumstances. U.S. schools have attempted to meet two important goals: enriching the classroom experience and providing access to education to a wide and varied population. In addition, these two issues, frequently referred to as "excellence and equity," have been accompanied by a host of other considerations.

State-mandated changes. State-mandated changes in curriculum call for more breadth and depth in courses that schools, particularly at the secondary level, are required to offer. These reforms affect schools of every town, city, county, and school district in America.

Requirements for high school graduation have been radically upgraded in many states, with special emphasis placed on mathematics, science, and languages. State colleges and universities across the country also are emphasizing the importance of these subjects by elevating admission requirements in these areas. Unfortunately, the task of meeting requirements at both the secondary and college levels is aggravated by a shortage of appropriate teachers and by budgetary pressures.

In addition to the curricular changes called for, most states now require teachers to participate in professional development or in-service training courses on a regular basis. The importance of professional development cannot be underestimated: To keep pace with the expanding educational requirements of their students, teachers must stay current with the most recent advances in their fields. For many teachers such courses are unavailable, inaccessible, or at best inconvenient.

Yet these schools must provide the basics of a good education and, if possible, broaden their students' intellectual exposure beyond the confines of their immediate locales. Struggling to provide a basic education to all students, many schools have few remaining resources with which to meet the unique needs of individuals who either have difficulty learning or are intellectually gifted. This is especially true for schools located in areas that are culturally isolated, economically disadvantaged, or both.

When resources must be stretched just to address the needs of the majority of students, gifted students, some of whom perhaps have the potential to provide signature insights about our world and its problems, may go unchallenged. This is a painful situation because, unless such students are challenged early, their ability to see unique relationships and to optimize their conceptualizing skills may be lost forever.

Teacher shortages are still another area of concern. The current shortage of qualified teachers in three key areas—math, science, and languages—is projected to worsen dramatically over the next two decades. For lack of a better alternative, some secondary schools have resorted to hiring teachers to teach subjects for which they are less than fully prepared. Finding teachers qualified to teach English as a second language is particularly critical in many locations.

Math and science teachers are sadly leaving the profession. In the 1999–2000 school year, to take one example, more than 45,000 math and science teachers left teaching after the school year. [103] This situation is causing schools to take measures unorthodox in education circles to attract and retain certified math and science teachers. In 2007, for instance, one county in North Carolina offered $10,000 signing bonuses for new Algebra teachers to sign up in their district. [104]

This problem is shared by both rural and urban schools. Often schools cannot afford the luxury of hiring teachers for courses such as trigonometry or Latin if only a few students will enroll. And some schools cannot convince subject-qualified teachers to relocate to their geographic area.

The ongoing dilemma of whether to focus financial and teaching resources on breadth or depth in the curriculum presents yet another problem for the world's schools. Struggling with budget and personnel constraints, many schools must choose between a curriculum that covers a large number of subjects lightly or an intensive, highly focused curriculum that covers key subjects in depth but other areas only superficially, if at all.

This dilemma cuts to the heart of the curriculum reform debate: Will a broad-based, general education or a more focused education (for example, a concentration on math and science) better prepare students for the world they will face as adults? Proponents for both sides of the debate have presented compelling arguments over the past several years. Most educators still believe the goal is to find a way to offer both breadth and depth, ensuring the most comprehensive educational grounding possible.

President George Bush's signature piece of bipartisan legislation No Child Left Behind (NCLB) aimed to bring greater accountability and measurability to the educational system. Aiming to make America more competitive, its goal was to have its students and teachers attain a level of proficiency. Critics have cited that NCLB resulted in a system that did not improve students' overall scores as compared against students in other countries and taught teachers to "teach to the test."

Former Secretary of Education Margaret Spellings stated that with NCLB, "we have shifted the national conversation. Instead of just asking how much we spend, we're finally asking whether students are learning-and we're holding ourselves accountable to change if they aren't." [105] Yet the fact remains, according to an article in *The Nation*, "the U.S. ranks twenty-eighth of forty countries in mathematics, right above Latvia, and graduates only about 75 percent of students, instead of the more than 95 percent now common elsewhere." [106]

Poor results like these are occurring in a country that spends more on education than most any country in the world. In the U.S. in 2005, the public expenditure on education as a percentage of GDP was 7.1 percent in 2005. [107] As *Time* magazine put it:

> "The U.S. spends more per pupil on elementary and high school education than most developed nations. Yet it is behind most of them in the math and science abilities of its children. Young Americans today are less likely than their parents were to finish high school. This is an issue that is warping the nation's economy and security." [108]

As the Secretary of Education Arne Duncan said, "America's economic security tomorrow is directly tied to the quality of education we provide today. This is our task. This is our challenge." [109]

Twenty-five years ago we were a nation at risk. Even after the immense amount of resources we have poured into the system, we are still a nation at risk. Why? Clearly, we need to embrace change.

CHAPTER THREE
The Roots of Cyberschools

"The data suggest that by 2019, about 50 percent of high school courses will be delivered online. In other words, within a few years, after a long period of incubation, the world is likely to begin flipping rapidly to student-centric online technology."

Clayton M. Christensen in Disrupting Class [110]

Cyberschools are a major part of the answer to the quest for affordable, accessible higher education. Delivered totally virtually, they are completely disruptive. In breakthrough, new ways, they employ almost every communications technology application developed, whether that be a TV course, courseware, online instruction, an Internet class, a virtual campus, or an e-book.

Cyberschools via electronic instruction won't solve all of the world's educational delivery problems, but they can help. If developed wisely and with the combined efforts of the public and private sectors throughout the world, distance education will use increasingly available communications technology to bring education to the learner. That learner can be in the U.S. or Europe or anywhere on the globe where the Internet is accessible.

Online education is attractive for four reasons in particular, as explained by the *College Student Journal*. Students with family or work constraints benefit from the convenience of cyberschools. Students who live in crowded cities no longer have to travel back and forth to campus. Individuals can choose from different online universities and have greater choice since location is not an issue. Finally, students benefit from their capacity to communicate globally and can get a broader perspective on various topics. [111]

Furthermore, they can be blended with traditional learning, so that students can have the best of both worlds: individualized training online and personal instruction where the student/teacher dynamics are fully realized. Plus they can offer more individualized training for the handicapped and greater access to AP courses in rural areas.

THE MEDIUM ISN'T THE MESSAGE

Much of my life has been devoted to the development and deployment of communications tools for the public at large. During that process, I have found that distance education is an instance in which the late media philosopher Marshall McLuhan's (www.marshallmcluhan.com) maxim, "The medium is the message," stated in his seminal work, *Understanding Media*, should not be taken literally. [112]

Though communications technology often fascinates us, the content it conveys — education, entertainment, news, and the like — was and still is king. Striving for human interaction to understand, interpret, and debate that content just happens to consume most of our lives. McLuhan's later work, *The Medium is the Message*, spelled out in more graphic terms how technology can shape our perception of content. [113] This undeniable phenomenon is the driving force behind much of the concern and debate over education and technology and is addressed with more detail in later chapters.

A BRIEF HISTORY

Historical context is useful in understanding the evolution of distance education from its text-based, correspondence course beginnings to its current foundation in technology. While some of the roots of cyberschools grow from distance education, they are in many other ways altogether different, new and dynamic. Cyberschools, enabled by the revolutionary growth of the Internet, offer a level of interaction and interactivity that is brand new — and completely transformative. We cannot fully understand the nature of cyberschools without first exploring how they originated in distance education and the rise of technology in the last two centuries.

Early examples of distance education are generally attributed to the late 19th century, when formal correspondence courses were developed. But the first distance learner to receive full university credit probably did so in the 18th century, when a homebound student on a remote agricultural estate made informal arrangements with a university lecturer to receive course notes and textbooks by mail and completed examinations in writing. The lecturer likely pocketed an "incentive" fee from the student, and university officials were never aware the student on the class roll was a phantom.

As social, demographic, and economic changes shaped both the 19th and 20th centuries, some leaders in education worked to fashion new ways to bring education to those who wanted it. One of the most outstanding higher education advancements in 19th century America was U.S. President Abraham Lincoln's 1862 signing of the Morrill Act, which initiated development of a system of state-supported universities intended to make college education more affordable and available to U.S. citizens. The act provided a 19th century bricks-and-mortar solution to the education distribution dilemma in one part of the world. The 1887 Hatch Act established agricultural experiment stations, followed in 1914 by the Smith-Lever Act, which authorized county extension agents for agriculture and home economics. These were some of the first attempts to take education directly from state universities to adult learners in the U.S. Today, arguably, the most successful efforts to bring higher education to more people at affordable costs involve distance education.

Initially, telecourses, or televised instruction, proved to be one of the most promising of the technology-based distance education alternatives. Advances in communications technologies such as cable television, fiber optics, microwave, wireless telecommunications systems, satellites, microcomputer networks, fax machines, videocassette recorders, and the Internet have allowed telecourse and courseware design and delivery to become even more effective. Now the explosion of the usage in mobile phones, PDAs, wikis, blogs, Second Life, Facebook, Twitter and electronic devices such as Amazon.com's Kindle and their probable progeny open up a whole host of new opportunities.

Beginning in the mid-1990s, the rapid evolution and adaptation of the Internet with its graphical and interactive World Wide Web provided an immediate distance learning medium that distance teaching institutions have been quick to employ. High-speed cable TV modems capable of delivering TV broadcast and Internet services, telephone lines with supercharged asymmetric digital subscriber line (ADSL) or integrated services digital network (ISDN) service, and satellite transmissions beamed directly to homes with telephone and Internet return links are now well established as viable infrastructure for delivery of distance education. The potential exists to turn every home on the globe into a real-time, interactive classroom.

WORLDWIDE DISTANCE HIGHER EDUCATION

Educational systems in sparsely populated countries such as Australia, Canada, and the Scandinavian countries have employed distance learning programs largely in the form of by-mail correspondence courses for more than 100-years. Distance education is used effectively in Australian higher education, in particular. Take, for example, the country's fastest growing higher education online service: Open Universities Australia (OUA). Over 110,000 people have already studied at OUA since its founding in 1993. Total unit enrollment has moved from just over 25,500 in 2004 to about 67,300 in 2007. Student numbers grew from 20,000 in 2006 to over 23,500 in 2007. While the majority of students are in Australia, courses are available globally. Undergraduate courses have no first year entry requirements, and there are no quotas for most courses, hence the "open" aspect. OUA is owned by seven leading universities: Curtin University of Technology, Griffith University, Macquarie University, Monash University, RMIT University, Swinburne University of Technology, and the University of South Australia. [114]

Countries are joining together across the world to make courses available for free over the Internet. The OpenCourseWare Consortium, for instance, is a collaboration of more than 200 higher education institutions and associated organizations from around the world. Member organizations come from countries as dispersed as China, South Africa, Australia and Columbia. [115]

In many countries traditional higher education institutions collaborate to design and deliver distance education. For example, Fédération Interuniversitaire de L'Enseignement à Distance in France, which coordinates distance education centers at 36 universities, had 30,000 students as of 2006.

In Ontario, Canada, Contact North serves 89 small and remote communities through coordinating centers in member colleges and universities. It offers through 13 Ontario colleges and universities over

200 courses and 80 part-time and full-time programs. Founded in 1986 by the Government of Ontario, it facilitated 12,970 course registrations in 2007–2008 in 698 credit full-time and part-time programs and courses. [116]

There also are stand-alone distance learning institutions in more than 20 countries. Many have huge student enrollments. Established in the early 1980s and offering more than 500 courses a year, China's Central Radio and Television University (CCRTVU) is the largest open-learning institution in the world. CCRTVU has an annual total enrollment of about 2.3 million students. [117]

Just look at a selection of the other distance learning institutions founded over nearly a 40-year time period. Other distance learning institutions, listed in the order in which they were founded, include: Open University of South Africa (1951), British Open University (1969), Universidad Nacional de Educación a Distancia, Spain (1972), Everyman's University, Israel (1974), Allama Iqbal Open University, Pakistan (1974), Universidad Estatal a Distancia, Costa Rica (1977), Universidad Nacional Abierta, Venezuela (1977), Anadolu University, Turkey (1981), Open Universiteit, The Netherlands (1981), Sri Lanka Institute of Distance Education (1981), Kyongi Open University, Korea (1982), University of the Air, Japan (1983), Universitas Terbuka, Indonesia (1984), Indira Gandhi National Open University, India (1985), National Open University, Taiwan (1987), Al-Quds Open University, Jordan (1987), Universidade Aberta, Portugal (1988), and the Open Learning Institute of Hong Kong (1989).

But it was the U.K.'s highly innovative British Open University (BOU) that quickly became an international distance education model by making college-level learning available to the general public. This institution truly opened the doors to distance learning and defied the "education for the elite" philosophy that still to a certain extent dominates European higher education systems.

Championed by then Prime Minister Harold Wilson, BOU was created as an alternative system for earning a higher education degree. Its courses blended print, radio, and some video presentations with campus visits. They were designed to appeal to students unable to attend universities full-time or in residence. There were no admission requirements. Anyone could enroll in BOU, but only students who successfully completed course requirements could obtain a degree.

To support its far-flung population, BOU established sites throughout the U.K. where students could take exams and meet with tutors. The British Open University, founded in 1969, is now a major force providing 21 percent of all higher education in the England. It has over 150,000 undergraduate students and more than 25,000 post graduate students. It has expanded throughout the world to Belgium, France, Greece, Hong Kong, Israel, Italy, Luxembourg, Malaysia, the Netherlands and Portugal. [118]

TV: EDUCATION DIRECT TO YOUR LIVING ROOM

Telecourses have been part of the U.S. educational delivery system since televised classes were first broadcast into America's homes more than 35-years ago. Typically received by ordinary home antennas from local broadcast television stations, these early, rudimentary telecourses brought traditional classroom presentations directly into students' living rooms.

Chicago Citywide College, an extension of the City College of Chicago, took the lead in testing and developing this new educational delivery system. [119] Supported by a grant from the Ford Foundation's Fund for the Advancement of Education, Chicago Citywide College began broadcasting telecourses over Chicago's public education television station, WTTW, in 1956. From those early days of trial-and-error experimentation, Chicago Citywide College has continued its commitment to expand the applicability and enhance the effectiveness of telecourse instruction.

And, although many other colleges have since followed its lead, Chicago Citywide's program is generally recognized to have set the stage for educational television today.

Also in 1956, while Chicago Citywide College was establishing itself, another Ford Foundation–supported project for television in higher education was launched at Pennsylvania State University. The purpose of the Penn State project was to explore the potential of closed-circuit television for on-campus instruction. It was successful. The project had produced 28 courses for the university by 1966.

Beyond closed-circuit courses, the 1950s spawned university-owned and-operated television stations. Examples include KUON, at the University of Nebraska–Lincoln, and WUNC, at the University of North Carolina at Chapel Hill. Many have since developed into highly effective statewide networks.

Widespread experimentation with telecourses continued at U.S. universities through the 1960s, '70s and '80s. Educators at Michigan State University, at East Lansing, American University in Washington, D.C., Case Western Reserve University in Cleveland, Ohio, and Iowa State University in Ames, among others, explored the possibilities offered by instructional television. They worked with a variety of professionals — teachers, instructional designers, graphic artists, educational technologists, and students — to find the most effective ways to create and deliver telecourses.

COMMERCIAL TV'S EARLY FORAYS INTO EDUCATION

Colleges and universities that did not own a station or a closed-circuit system got their opportunity to experiment with television courses when the commercial networks became interested in educational television. WCBS/New York first broadcast New York University's Sunrise Semester series on comparative literature in 1957. By 1958, NBC was broadcasting "Atomic Age Physics" on Continental Classroom over

150 network stations across the country. Funded in part by another grant from the Ford Foundation, the physics series received high marks from educators for academic quality and the usefulness of the accompanying support materials for students and local teachers. More than 300 colleges and universities offered "Atomic Age Physics" the first year, and several other courses followed in succeeding years.

Unfortunately, the series required a heavy subsidy, and NBC dropped it after a few seasons. Nevertheless, educators, programmers, and producers learned valuable lessons about telecourses from the experience. Indeed, they learned that a program created with high academic standards would be accepted by teachers and students; that a market for such programs existed; and that, as always, financial issues needed to be considered. After a 25-year run, NYU's Sunrise Semester was discontinued in 1982.

PUBLIC TELEVISION

Another landmark event for telecourses in the early 1960s was the passage of the Federal Educational Television Facilities Act of 1962. This legislation empowered the federal government to fund the building and equipping of public television stations, thereby extending educational television's broadcast reach.

In response to the growing interest in telecourses, the Great Plains Regional Instructional Library was created in 1963 by an agency of the KUON-TV/Nebraska Educational Television Network in affiliation with the University of Nebraska–Lincoln. The library's goal was to serve as a clearing-house that would acquire, maintain, and lend to schools those programs and series that had continuing educational value.

Now called the Great Plains National Instructional Library (GPN), it is the largest nonprofit distributor in the instructional/educational market. GPN distributes a wide variety of video programs ranging from an Emmy award-winning PBS children's series through college-credit speech communication telecourses.

WATERSHED: THE PUBLIC BROADCASTING ACT OF 1967

Probably the biggest attempt to advance educational television in the U.S. occurred with the passage of the Public Broadcasting Act of 1967. It recognized the potential of broadcast television to inform and enlighten as well as entertain the public. This legislation authorized the creation of the Corporation for Public Broadcasting (CPB), which was charged with the "responsibility of assisting new stations in getting on the air, establishing one or more systems of interconnection, obtaining grants from federal and other sources, providing funds to support local programming, and conducting research and training projects." [120]

The CPB was not a production or networking facility. PBS was created in 1969 to serve as CPB's television network. Its functions were to select, schedule, and distribute programming for the widespread system of PBS-affiliated stations. Through that network, a nationwide system of public television comprising some 350 local stations came into being.

COMMUNITY COLLEGE TV

In the mid-1970s, community colleges began producing their own tele-course series and related support materials to attract broader audiences and extend the reach of their campuses. Miami–Dade Community College in Miami, Florida, Coastline Community College District in Fountain Valley, California, and Dallas County Community College in Dallas, Texas, were three of the colleges most active in this field. Since then, many community colleges have joined together in regional consortiums that currently produce some of the best telecourse programs available throughout the world.

PROGNOSTICATION

All of the foregoing have helped to form the root system of cyber-schools. They all share the same soil. However, there is reason to believe that the advent of cyberschools may, even over a short span of time, prove to be the most compelling tool of them all.

CHAPTER FOUR
The Virtual Classroom of the 21ˢᵗ Century

"The Virtual Classroom is one of those things that is best experienced, like a sunset or a swim in ocean waves, in order to fully understand it."[121]

Starr Roxanne Hiltz in The Virtual Classroom

The days when the overhead projector was the highest-tech teaching tool to grace the halls of ivy are over. The numbers prove it. For example, in 1994, email was used in only eight percent of college courses in the U.S. By 1999, that percentage had soared to 54 percent. By 2002, 72 percent of college students checked their email at least once a day. [122] 85 percent of teens ages 12 to 17 engage at least occasionally in some form of electronic personal communication, which includes text messaging, sending email or instant messages, or posting comments on social networking sites, according to the Pew Internet & American Life Project. [123] In 1996, only 9.2 percent of college courses had Web pages. In 1999, the number had almost tripled to 28.1 percent. A modest beginning, but, as Kenneth Green, director of the U.S. Annual Campus Computing Survey, wrote in an issue of *Change* magazine, "Most colleges and universities have finally passed the point of critical mass affecting the instructional use of information technology." By 2008, wireless access is prevalent throughout college campuses. Overall, about two thirds (67.6 percent) of classrooms have access to wireless, up from 60.1 percent in 2007 and 31 percent in 2004. [124]

Technology doesn't necessarily have to come in the form of a microcomputer. For technophobic teachers, perhaps just an Internet access device would suffice. But the higher education establishment must adapt to technology in the lecture hall, get used to technology as the lecture hall or even forget lecture halls. A true paradigm shift needs to occur in the way college educators and the higher education establishment perceive and design the education environment.

While over half of the adult Internet population is between 18 and 44 years old, greater percentages of older generations are online now than in the past, according to the Pew Research Center's Internet & American Life Project surveys from 2006–2008. "The biggest increase in Internet use since 2005 can be seen in the 70 to 75-year old age group. While just over one-fourth (26 percent) of 70 to 75-year olds were online in 2005," 45 percent of that age group is now online. [125]

Equally important, technology can be used to create a less expensive way to deliver higher education to those who want it. Although not the same as the experience of the college campus, advances in computers, cable television, satellite and mobile technology and the ubiquity of the Internet will make higher education available to the most people at the lowest cost — worldwide.

In 1996 in the first edition of this book, I predicted that the dawning of the 21st century would usher in the full use of technology-based institutions that could function side by side with traditional universities. This has come about as a number of universities around the world—including MIT, the University of Colorado, and many others—have gone online with a variety of degree and certificate programs that offer the same credentials to their graduates as their campus-based counterparts. [126]

THE HOME AS A CLASSROOM

The idea is to deliver education to people, instead of people to education. Why now? Because, as we all know, earning a living in post-industrial, knowledge-age society requires lifelong learning, training, and retraining at every level. For the vast majority, interrupting work life to study in a traditional university setting is out of the question.

What does "virtual" really mean? The term virtual is used in computer science to refer to something whose existence is simulated with software, rather than actually existing in some physical form. The virtual university, or cyberschool, is education dispensed from an electronic platform, instead of a lecture hall podium. Indeed, work on virtual classrooms and virtual universities has been taking place since the 1980s, both in the private sector and in the public sector at university consortia, governors' conferences, and public television studios. The rapid evolution of new software standards, and particularly the recognition of the World Wide Web as an open software platform, have led to major shakeups and shifts in the development of educational delivery software and standards by major technology companies.

Hundreds of companies across the world now offer helpful online tools, including videoconferencing. Two of the most sophisticated systems were developed by Hewlett Packard and Cisco. Cisco's high-definition TelePresence System, introduced in 2006, allows for near life-like and life-sized conferences to be held in real-time. The resolution and quality of the virtual conferences is stunning. While its price and required bandwidth are high, larger organizations are finding the savings in travel costs more than pays for itself. And the pricing of such systems, like most new technologies, will likely decline. Cisco is now encouraging University researchers to connect and interact. In October 2008, for instance, Cisco announced its funding of the Cisco TelePresence University Connection. The network began with 12 universities including MIT, Stanford University, and the University of North Carolina at Chapel Hill. Hewlett Packard also offers the competing Halo Telepresence System.

Apple's iChat AV and the iSight camera are just two examples of the many services available in the market to allow for desktop videoconferencing. Furthermore, many students view podcasting in higher education as an effective form of mobile learning. According to a study done by *Computers and Education*, 200 first-level students at a school were given podcasts after completing a course in Information and Communications Technology. Students found podcasts to be more effective revision tools than their own textbooks, the study found, and the podcasts were more helpful than their own notes in helping them learn. [127]

Abilene Christian University has even gone so far as offering each new student a digital media device, either an iPhone or an iPod. Students can receive alerts about homework, answer in-class surveys and quizzes, receive directions to their professors' offices, and check their meal and account balances, among more than 15 other useful web applications already developed. [128]

The world-renowned British Open University (BOU) in the U.K. used Sun's Java software to develop a program called Stadium, which is designed to allow thousands of students at a time to listen to and participate in special guest lectures given on the university's virtual campus. Stadium's creators tout it as an experiment in very large telepresence. Telepresence captures the mood of an event—applause, laughter, shouting, whispers between neighbors, and the like—using software. BOU is also making millennium-level shifts in its strategies by moving more of its courses from television delivery to CD-ROMs and the Internet.

Jones International University in 2006 launched its "cyberdome," a unique virtual arena that will serve as, among other things, the location for its university graduation ceremonies. The cyberdome's inaugural ceremony was the commencement of JIU's Class of 2005–2006.

EMERGING TECHNOLOGIES IN HIGHER ED E-LEARNING

The 2008 *Horizon Report,* jointly produced by the New Media Consortium and the EDUCAUSE Learning Initiatives, identifies six major technology topics to be integrated into learning in the next six years.

> "The two technologies placed on the first adoption horizon... grassroots video and collaboration webs, are already in use on many campuses. Examples of these are not difficult to find. Applications of mobile broadband and data mashups, both on the mid-term horizon, are evident in organizations at the leading edge of technology adoption, and are beginning to appear at many institutions. Educational uses of the two topics on the far-term horizon, collective intelligence and social operating systems, are understandably rarer; however, there are examples in the worlds of commerce, industry and entertainment that hint at coming use in academia within four to five years." [129]

VIRTUAL CLASSROOMS: BETTER THAN REAL?

For the uninitiated the concept is a little tough to grasp. But, as virtual classroom design innovator Starr Roxanne Hiltz, the Distinguished Professor, Emerita at the New Jersey Institute of Technology, advised,

> "The Virtual Classroom is one of those things that is best experienced, like a sunset or a swim in ocean waves, in order to fully understand it... think of all the different

kinds of learning tools and spaces and ritualized forms of interaction that take place within a traditional classroom, and within an entire college campus or high school. All of these things exist within a Virtual Classroom, too, except that all of the activities and interactions are mediated by computer software, rather than by face-to-face interaction." [130]

First introduced in the mid-1980s, Hiltz's technology was used by the ConnectEd, or Connected Education, program, part of the New School University in New York City, to link instructors in North America with students in Asia, Europe, and Latin America as well as at Upsala College (a small private liberal arts college) and the New Jersey Institute of Technology. [131] The technology was used much like an Internet dial up is used today: A student in Asia or anywhere with telephone access could dial a local telephone number to connect with the mainframe computer in New Jersey to receive any course material stored there, to leave or receive papers, or to communicate with an instructor or other class members. [132]

Another way to think about the virtual classroom is to compare it with how interaction takes place in a traditional classroom. In traditional classrooms, most interaction takes place via speaking, listening, reading, and writing. In the virtual classroom, interaction currently takes place by typing and reading from a computer. If that seems like a less than superior alternative to the traditional classroom, it really is not. Hiltz contended a collaborative learning environment that is computer mediated can support some activities that are difficult or impossible to conduct in face-to-face environments, particularly if there is a large class. In the virtual classroom, discussion and communication about the course become a continuous activity. If a student has an idea or a question, it can be communicated while it is fresh.

The advantages of the virtual university's being virtually anywhere — in a living room, a kitchen corner, or the local library — are obvious, but I'll reiterate them anyway: All students in a particular class don't have to be at the same place at the same time, week after week, for the course of a semester; and the virtual university is open 24-hours a day, seven days a week. For adult learners with jobs and family responsibilities, the ability to do coursework on their own schedules may make the difference between successfully completing a degree program and dropping out.

If you think about using traditional terms, in the virtual university interaction "spaces" are created within a software package and are used as so-called classrooms, where teachers lecture and where group discussions take place; there is a communication structure like office hours, through which student and teacher can communicate privately. The software also has the ability to administer, collect, and grade tests or assignments and the ability to divide a larger class into smaller working or peer groups for collaborative assignments. [133]

All this happens within a computer-mediated communication system. Two examples are the local area network, through which students and teachers can communicate via electronic bulletin boards, email, private news groups on the Internet (also called listservs), and private chat rooms. The local area network by its nature can restrict the virtual classroom to a certain defined area, like a college campus, and a course delivered over the Internet's World Wide Web can have a similar structure with password restrictions. And as technology rapidly evolves and the learning environment becomes more media rich, the possibilities are breathtaking.

The concept has been tested and is successful. In 2008, 26 states in the U.S. had state virtual school programs, and more than 50 percent of all school districts in the U.S. offered online courses to students in K–12 education, according to iNACOL.

In 2009 the U.S. Department of Education issued a report on their findings based on a survey of 51 independent studies of online education as compared to traditional face-to-face instruction. To quote the report, *Evaluation of Evidence-Based Practices in Online Learning – A Meta-Analysis and Review of Online Learning Studies*, "Students who took all or part of their class online performed better, on average, than those taking the same course through traditional face-to-face instruction." [134]

Clearly, the evidence suggests that online learning has a significant presence in the process of education.

VIRTUAL EDUCATION VISIONS

Visions of virtual education are increasingly "blended learning" visions. "My sense is that our concept of [online learning] is either/or: It is either in a classroom or it is online," said Bill Tucker, Chief Operating Officer of Education Sector, an independent education-policy think tank. "In the future, it is going to be pervasive, so we will not really know the difference… there is going to be such a multiplicity of choices, you will see all shades of online coming into offline, and they are going to mix and match so you do not know the difference anymore." [135]

John Watson, Founder of Evergreen Consulting Associates, sees four categories of online learning:

- State-led and state-funded programs such as the Florida Virtual; School (FLVS), the largest virtual public school in the country;
- Charter schools that are fully, or partly, online;
- District programs; and
- Consortium or network programs. [136]

In the 2008–09 school year, at FLVS, about 84,000 students will complete 168,000 half-credit courses, a 10-fold increase since 2002–2003. [137] FLVS is highly innovative. It only receives funding, as *Threshold Magazine* points out, for students who pass classes. Julie Young, FLVS President and C.E.O., put it simply, "There was so much skepticism around online learning that they wanted to 'put money where their mouth was.' Students don't succeed; we don't get paid." [138]

Multiple visions are transforming higher education. Western Governors' University, founded by 19 governors of the Western states and supported by over 20 major corporations and foundations, was an early prime example of a vision for a cyberschool.

In early 2000, it offered five technology-related degree programs and courses from another 47 higher education institutions. The governors first discussed their virtual university idea at the Western Governors' Association's annual meeting in June 1995 in Park City, Utah. Money, as often is the case in both government and the private sector, was the driver. They recognized the favorable economics of the cyberschool model. The concept grew out of a discussion about how to contain the costs of producing expensive distance learning courses and the states' limited capacity to fund increasingly expensive traditional higher education. What evolved was a vision for a combination of technology and face-to-face modes — specifically, cable television, Internet-based courses, and summer seminars — delivering education to far-flung students in all the idyllic nooks and crannies of this sparsely populated region.

The governors' objectives mirror those of many other distance learning programs. Their goals are similar, too, in that they mention extending educational opportunities to more citizens, reducing the costs of higher education, and shifting the focus of education away from "seat time" and toward competence. [139] WGU is the only university that has been granted regional accreditation by at least four regional accrediting commissions. WGU is also the first online university to be competency-based.

Today, the online university boasts over 10,000 students and is still growing rapidly. WGU students are scattered across all 50 states, range in age from 16 to 72, and pursue one of WGU's 42 online degrees or certificates in business, education, information technology, and health professions.

In Dr. Curtis Bonk's amazing new book entitled *The World Is Open*, he explains how web technology is revolutionizing education. [140] Dr. Bonk sets forth the ten world "openers" that evolve out of three converging macro trends, which are:

I. The availability of tools and infrastructure for learning (the pipes);

II. The availability of free and open educational content and resources (the pages); and

III. A movement toward a culture of open access to information, international collaboration, and global sharing (a participatory learning culture).

THE VIRTUAL LIBRARY

We know that knowledge is power. Dating back to the end of the Middle Ages, at the heart of every traditional university has been a great library. Libraries are essential elements in the process we call education.

Indeed, in the early 17ᵗʰ century, Sir Francis Bacon may have created one of the first library cataloguing systems. As part of his essay *In Advancement of Learning*, he divided "all knowledge" according to the faculties of memory, imagination, and reason. Thomas Jefferson used the same three categories to organize his own large personal library out of which grew the Library of Congress, which has become the largest repository of information in the world. In the knowledge age, we have taken the vision of a great library even further: At the heart of any great virtual university has to be a great virtual library. There is enough digital and digitized information today to develop virtual libraries filled with the world's knowledge; and it is available to anyone with access to a computer and a modem. Think about it: online access to the globe's important information by every scholar in the world.

Almost every major university research library in the U.S. has initiated a program to place part of its collections and archives into digital format, usually available for free on the World Wide Web. A similar spontaneous movement is taking place in Europe and parts of Asia, as evidenced by the Treasures from Europe's National Libraries collection on the Web, the Bibliothèque Nationale de France's program to digitize thousands of French texts and images, and the University of Adelaide's Electronic Text Collection. Much scientific research is now conducted online by widely scattered "virtual communities," and books are being widely digitized though there is controversy over how much will be finally accessible online and at what cost.

On June 6, 2005, in remarks to the Plenary Session of the U.S. National Commission for UNESCO, the Library of Congress' outstanding Librarian James H. Billington proposed the development of the World Digital Library (WDL). The following year, Dr. Billington explained that, "the goal is to create an online encyclopedia, freely available over the Internet, of important and interesting cultural objects from the world's countries and civilizations. These works will reside in a large, online repository that can be searched and used in different ways by teachers, librarians, scholars, and the general public." [141]

The Library of Congress inaugurated in April 2009 the beginnings of the WDL at UNESCO headquarters in Paris with items online representing all its 192 countries. Principal founding partners included the Alexandria Library; the Brown University and Yale University libraries; the Central Library, Qatar Foundation; the King Abdullah University of Science and Technology in Saudi Arabia; the National Libraries of Brazil, Egypt, China, Iraq, Israel, Russia, Serbia and Sweden; the University of Pretoria, South Africa; and the new Yeltsin Presidential Library in St. Petersburg Russia. Viewers can make comparisons across countries and timelines with curatorial commentary in seven languages.

Other digitized resources are already available. The Library of Congress's National Digital Library Program had placed 15.3 million primary documents of American history and culture online free of charge by the end of 2008. This "American Memory" web site has benefited from bipartisan support from the Congress and from major private philanthropic and corporate contributions from John Kluge, companies associated with John Malone, the David and Lucile Packard Foundation, the Kellogg Foundation, Ameritech, Bell Atlantic Corporation, the McCormick Tribune Foundation, The Discovery Channel, Eastman Kodak Company, Thomson Reuters Corporation, Compaq Computer Corporation, The Hearst Foundation, R.R. Donnelly & Sons, NYNEX Corporation, Nortel, Jones International University and others.

A separate ambitious venture is the Internet Archive. Founded by Brewster Kahle in 1996, it is a nonprofit archive offering permanent access for researchers, historians, and scholars to historical collections existing in digital format. At the end of 2008, the Internet Archive housed over three petabytes of information, according to Sun Microsystems, and is expected to grow at approximately 100 terabytes a month. It also features "The Wayback Machine" — a digital time capsule that allows users to see archived versions of Web pages across time. [142]

THE DECADE 'GOOGLE' BECAME A VERB

Since the last update of this book in 2002, one company has moved this vision of universal access to knowledge to new heights, Google. The company has lapped most competitors on the information superhighway raceway with some speed bumps in areas such as copyright and privacy. [143] Yet Google is reshaping entire industries. Whether it be newspapers, advertising, energy or healthcare, how each industry adapts in this new era could well determine its survival. Jeff Jarvis in his provocative book *What Would Google Do*, wrote as follows:

"There are two ways to attack the problems of these
industries: to reform the incumbents or to destroy them...
the wise course is to destroy your own models before
some kid in a garage — or in a Harvard or Stanford dorm
room — figures out a way to do it for you. Think like
Google, succeed like Google — before Google does." [144]

"Even before we started Google, we dreamed of making the incredible breadth of information that librarians so lovingly organize searchable online," said Larry Page, Google's co-founder. [145] The expansion of the Google Print project allowed Web surfers to download books in the public domain in their entirety and see three or four lines from copyrighted books.

Google's ascendancy is nothing short of remarkable. Consider: up to 30 million photos are uploaded to Google every day, and almost all of them are people; and about 10 hours of video are uploaded every minute on YouTube. [146]

The storage capacity benefits Google's YouTube. By 2019, it is expected that an IPOD-like device will be able to have 85-years of video on it. "You can't watch it your whole life. You're dead before you finish watching all the videos on this device," joked Eric Schmidt, CEO of Google, speaking at the Economic Club of Washington DC in June, 2008. [147]

So what's the next wave for Google, what are the new products that most excite the CEO of Google? "The ones that are most impressive are the ones that use artificial intelligence to do things that I cannot imagine are doable... [and] the automatic translation that is being done now," Eric Schmidt told The New Yorker's Ken Auletta in 2008. "We will eventually do 100... languages where you can automatically translate languages into another. And the translations are done by computer, not done with humans. That alone will have a phenomenal impact." The second example Schmidt cited relates to "geo positioning... Think about all the places that have never had reliable maps." They have engineers developing maps of cities that have never had maps before. [148]

Google offers untold opportunities for cyberschools and those who take education courses. Jeff Jarvis added,

> "Who needs a university when we have Google? All the world's digital knowledge is available at a search. We can connect those who want to know with those who know. We can link students to the best teachers for them (who may be fellow students). We can find experts on any topic... Call me a utopian but I imagine a new educational ecology where students may take courses from anywhere and instructors may select any students, where courses are collaborative and public, where creativity is nurtured as Google nurtures it." [149]

It is hoped that as more and more rare, classical literary artifacts become digitized and available to living room–based patrons of virtual libraries, educational barriers that divide societies will begin to crumble. I believe these libraries in cyberspace will become to the 21ˢᵗ century what the fax machine was to the late 20ᵗʰ century. Information disseminated by fax helped change the world's political balance when the Iron Curtain collapsed in the late 1980s. Wouldn't it be wonderful if in the 21ˢᵗ century the ability to share the world's knowledge through virtual libraries and global virtual universities brought the world to the next level of understanding? It can happen. Just realize that the products in the 21ˢᵗ century will be entirely different than those in the past given the evolutionary speed of technology.

THE TRANSFORMATION OF TIME ITSELF

Time itself is on the side of the virtual classroom in more ways than one. First, cyberschools are not dictated by 50-minute time increments. Virtual classrooms are open 24–7 and can be accessed at any time. For the busy professional whose major scarcity is time, this convenience is a major factor for choosing virtual education. Students don't need to waste the time of travel, or idle conversation. They can go right to the heart of the subject matter and stop and start as they please. In the non-synchronous space of most cyberschools, the world has only one time zone.

TEACHER HOLOGRAMS

Despite the evidence, there are arguments in traditional educational circles that virtual universities, classrooms, and libraries lack the synergism of the traditional delivery systems and that students will, therefore, learn less. Understandably, professors appearing as holograms is a discomfiting idea. I can understand why these entrenched opinions exist, but I urge their proponents to rethink them.

Here's why. In the virtual classroom, it is nearly impossible to be a passive learner. Students nearly always must react or provide some appropriate input in order to continue to the next phase of an assignment. In addition, virtual courses are more often designed to be collaborative efforts among students than are the traditional lecture hall courses we all knew and loved — at least for the sheer anonymity they provided.

All mediums of communication have their advantages and disadvantages. But the research really does show there is no significant difference in the student's ability to learn using technology-based educational tools — and not just for computer-aided teaching. Long ago, research into whether television was an inferior learning tool proved there is no real difference between learning from TV and learning in the traditional classroom. [150]

Researchers and academicians debate the pros and cons of virtual classrooms, but most early indicators are that the same is true for education via computers and the Internet. At the beginning of any paradigm shift—and the knowledge age is demanding a paradigm shift in educational delivery—there must be explorers. The technology available today plus the current economics of higher education demand that we either become adventurers ourselves or provide the opportunity for others to be.

CHAPTER FIVE
The Internet Comes to School

*"With the development of the Internet,
and with the increasing pervasiveness of
communication between networked computers,
we are in the middle of the most transforming
technological event since the capture of fire."*

John Perry Barlow in Harper's Magazine [151]

In the mid-1990s, arguments raged among faculty, school administrators, and Internet proponents about whether the Internet should be a formal part of education's infrastructure. For those who resolved the argument in favor of the Net, another formidable barrier had to be confronted: where would the money come from to provide technology access in cash-strapped schools and university systems around the world?

Then an amazing phenomenon took hold. Students of all ages began bringing the Internet to school in the form of Web-delivered research and the results of email chats with their peers about their assignments. At the same time, both private industry and governments began designating contributions and budgets to provide Internet access and more computers for classrooms. This coincided with a decrease in the cost of Internet hardware as the price of computing chips dropped to an all-time low. By late 1998, the price of an individual computer workstation dropped to less than a thousand dollars, down from two to three thousand only a few months before. By 2008, the price of some computers fell below $500; and some Netbooks were even under $250. Additionally, the amount of bandwidth available has greatly increased and continues to do so. Cell phones, too, continue to evolve rapidly.

In addition, a whole range of new tools are now available for students. Ranging from an explosion of avenues of communications (such as blogs and wikis) and social media such as Twitter, MySpace and Facebook, students have an unprecedented number of ways to communicate with each other, to learn and to engage with their teacher.

Cable in the Classroom (CIC), the cable industry's education foundation launched in 1989, has played a central role in the expansion of Web-delivered learning. Cable operators have provided free cable and broadband connections to more than 81,000 schools and libraries around the country. Those connections bring access to more than 500 hours of educational television programming each month and countless online resources from cable networks. Over the past 20 years, the cable industry, through the Cable in the Classroom initiative, has delivered everything from distance learning and electronic field trips to online courses, webcasts, and interactive digital learning modules, said Frank Gallagher, Director, Education and Media Literacy, Cable in the Classroom.

Innovative computer uses on the part of students, along with industry and government initiatives and the economics of inexpensive micro-computing, launched an enormous thrust forward that completely overwhelmed the Internet's detractors within the educational establishment. In a process characterized by its near-perfect symmetry, a technology born in the research labs of major government and university research institutions was opened to vast consumer markets and then flowed back into education through a multipronged approach that made it far more valuable.

Between 1996 and 1999, the number of U.S. K–12 schools with Internet access increased from 32 percent to more than 90 percent. In early 2000, connectivity in classrooms moved ahead in other countries as well. By 2007, an extensive survey by Sloan-C, the non-profit group that studied trends in online learning, found that nearly two thirds of all districts surveyed have students taking either online or blended courses with another 21 percent planning to introduce them over the next three years. [152]

THE WEB WHIRLPOOL

At the end of 1994, *Business Week* touted the World Wide Web as "the hippest, most exciting neighborhood on the Internet." [153] In the 21st century, the Web is beyond hip; it's mainstream. Somewhere in the graphics of nearly every television commercial or promotional announcement, there's a Web address. Even if they aren't "wired," most people in North America, Europe, many parts of Asia and South America, and some parts of Africa know that www.something.com or .org is an Internet Web address.

In a true testament to the Internet's cultural acceptance, Hollywood immortalized it on celluloid in the 1995 film *The Net*, and that was barely the beginning. From futuristic high-tech thrillers such as *The Matrix* and *Independence Day* (in which computer viruses were fed to alien invaders) to traditional romantic comedies such as *You've Got*

Mail, the wired world has been adopted by the celluloid camp. If not the new darlings of entertainment, at the very least the Internet and computers are ready-made plot devices that no gimmick-minded screenwriter or director can pass up.

Louis Platt, the former CEO of Hewlett-Packard, called the Internet the newest utility, right up there with water, power, and telephone service. In fact, the Internet is converging with the utility industry as gas and electric companies string fiber-optic lines through gas pipelines and send bits and bytes along high-transmission electricity lines. Andrew Laursen, a Senior Vice President of Product Development at Scalent Systems, said when he was at Oracle, "The Internet will subsume television, radio, and retail. The Internet will be everything."[154] The Internet is like a whirlpool that may eventually absorb cable and many other industries.

Cable companies were instrumental in allowing ordinary Americans to receive broadband Internet connections at unprecedented speeds and at an affordable cost. Verizon and its Fios system are also providing strong broadband access to the Internet.

In distance education, the Internet truly is a transforming technology. Scholars Chester Finn and Bruno Manno contended that "even if we have only the PC and the Internet, we have enough to revolutionize education."[155] Through electronic delivery, it is possible to shift the emphasis in education from the institution to the student, where it belongs. And it is possible to deliver education more cheaply. As I've mentioned, it is becoming more apparent every day that fiscal concerns will have much to do with shaping education, its quality, and how it is delivered.

We now have the ability to organize information and manage it like we've never been able to before. Organizing the world's information has become, in fact, the mission of Google; founded in 1996, it now has the dominant world market share in search.

We also have a critical need to lower the cost of education. We must strengthen educational institutions—it's in everyone's interest—and at the same time realize that the notion of educational institutions as physical places—real estate, if you will—is diminishing. Everyone in higher education will be affected by the Internet, including college presidents, regents, faculty members and students. The Internet allows smaller classes and larger classes, as well as close one-on-one interaction with students via electronic chat rooms, email, etc. Finn and Manno envisioned an online educational resource in which course lectures are available not in 50-minute chunks, but in two to five-minute video segments closely matched with paragraphs in a textbook, and a video of an expensive-to-duplicate demonstration, with problem sets right at hand. [156]

THE FUTURE IS HERE

Manno and Finn's vision is already a functional reality. Currently, many distance learning certificate and degree programs offer courses via the Internet and World Wide Web that use combinations of video streaming, email, online chat groups, and text-based materials. There is a book, although now it's dated, called *The Internet University* devoted entirely to cataloguing Internet-based higher education worldwide and a quick browse through the Web reveals online public and private university offerings galore. [157]

Transformed from a concept that was considered esoteric and a hard sell in the mid-1990s, online course offerings are now almost commonplace, though the quality of many reflects the need for customers to choose wisely and judiciously. Programs that are accredited attract most of the students, but online diploma mills proliferate and prey upon the impatient and unwary.

In the U.S. and Canada, dozens of schools offer online courses. A visit to the web site GetEducated.com (www.geteducated.com) provides a current, if not complete, sampling of the many online degree programs, including over 80 undergraduate and graduate program degrees offered in the U.S. and Canada.

Among the offerings listed there are bachelor's or master's degrees in:

- distance learning from Athabasca University, Canada's Open University;
- healthcare, paralegal studies or project management from George Washington University;
- applied economics from Georgia Southern University;
- business administration and business communications from Jones International University;
- supply chain management, geographic information systems, earth science and homeland security and public health preparedness from Pennsylvania State University; and
- science in health informatics from the University of Illinois at Chicago.

Web sites listing bachelor's degree programs, degree completion programs (usually for students who have completed two or more years of college coursework), and numerous certificate and personal skill programs now proliferate. The biggest challenge for the student is gauging which ones are of high quality and affordable.

The University of London and the University of South Africa were listed among the world's top three online education programs, and Jones International University was listed as the sixth highest ranked by the Global Academy Online. [158] Consider these success stories in online education:

- Pennsylvania State University's virtual classroom offers over 500 courses, 15 undergraduate degrees and 11 graduate degrees and has over 7400 students enrolled. Long a leader in distance education, Penn State has used email in its independent study courses since early 1995 and now fully uses Web delivery; and

- The online pioneer University of Phoenix, founded as a site-based university in 1976, is now the largest private university in North America with over 100 degree programs at the associate's, bachelor's, master's and doctoral levels in more than 200 locations. As of February 2008, site based and online combined degree enrollment was astounding: over 330,000 students. The average student age is 35, and two thirds of the students are female. [159] In 2008, it was ranked 2nd in the list of the world's top 10 online universities (only the University of London surpassed it). [160] The ranking group Global Academy Online called Phoenix "a model copied wide and far by traditional schools." [161]

The cost difference between attending a face-to-face college or university and the online university can be substantial. While total tuition and fees to attend a four year college cost $22,000 per year on average in 2005, they cost about $12,000 at the University of Phoenix. [162] Nearly 3.5 million students were taking online courses in the Fall of 2006, according to the Sloan-C. That was up nearly 10 percent from the previous year. In fact, by 2008, more than two thirds of educational institutions offered some form of online learning. [163] The biggest growth in online learning has been at two-year colleges.

THE WEB 2.0 EXPLOSION

Brick and mortar and virtual educational institutions are finding more and more intriguing and inventive ways of integrating Web 2.0 into their curriculum. Consider these examples cited in *The Horizons Report*, produced by The New Media Consortium and the EDUCAUSE Learning Initiative:

- Courses from UC Berkeley are available on a specially branded YouTube channel;
- UMBCtube, a custom YouTube channel for the University of Maryland Baltimore County, enables the campus to blend community generated content with video. It is designed to complement its main course media portal on iTunes U [164]; and

- A course on Digital Entrepreneurship at The Rochester Institute of Technology created a social networking Ning network bringing undergrads into contact with "over a hundred graduate students, venture capitalists, faculty, practitioners and business owners around the world." [165]

WORLDWIDE LEARNING REVOLUTION

The online learning revolution, as described earlier, is not confined to the U.S.

Online education is still nascent but growing fast in two of the world's largest countries: India and China. The online education market in India in 2008 is estimated at generating $200 million in revenue, and industry experts predict it will reach $1 billion by the end of the decade. "Online education addresses some of India's shortcomings: a dismal education system, limited reach, and a severe paucity of faculty," according to *Business Week*. [166]

In China, online courses currently supplement some K–12 traditional classroom learning. Three major obstacles are hampering growth in online learning in China, according to an iNACOL report in 2006: a lack of qualified candidates with the appropriate certifications, the difficulty selling the new concept of online learning into the traditional educational system and the poor Internet technology. [167] Despite these significant obstacles, 67 universities were participating in 2006 in a pilot to run online programs in both large and mid-sized cities in China. In 1999, according to iNACOL, the online educational market in China amounted to 1.7 billion US dollars, while in 2004 it rose to 23.1 billion US dollars. [168]

The British Open University has around 150,000 undergraduate and more than 30,000 postgraduate students. More than 25,000 of its students live outside the U.K. One setback it faced earlier in the 1990's provides some valuable lessons learned. As early as 1993, British Open

University was contemplating expanding its services by creating an electronic campus. It followed through with these aspirations in the Spring of 1999 when it opened a sister campus program in the U.S., but it ceased operations in June 2002.

In a post-mortem assessing why the campus closed, an article in the respected *Educause Quarterly* cited five factors:

1. The loss of an important advocate and diminishing support from its parent institution;
2. Conflicts with the OU's established curriculum [leading to the need for the adaptation of some courses for a U.S. market at significant cost];
3. Challenges in entering a new market, particularly the competition from 3,885 institutions of higher learning;
4. Lack of accreditation; and
5. Problems with business planning. [169]

A top official reportedly said, "the biggest mistake we made was getting started with undergraduate education. The [Open University] MBA is one of the largest and most highly regarded in Europe... We should have done an MBA or Americanized the OU MBA first." He did believe USOU would have made money eventually. [170]

Despite some of the setbacks that most all online ventures faced during the dot com meltdown, the growth internationally continued unabated. In Latin America and other developing areas of the world, online and other electronic educational options arguably are the most cost-effective way to educate enough of the population to bring those countries into the 21st Century knowledge age. Inflation and interest on huge foreign debts had a profound impact on educational funding in Latin America.

Whitney International University System, a Bermuda-based company with management offices in Dallas, has acquired controlling stakes in four institutions: Brazil's University Center Jorge Amado, Panama's

Isthmus University, and two community colleges in Colombia. Through a number of other alliances, it expects to enroll over 40,000 students in distance-learning classes in Latin America in 2008 and to double that number by the end of 2009. Tuition, while still high, is still about half of what on campus students pay. [171]

The Fundación Cisneros supports the world's first pan-regional educational television channel, cl@se, that delivers quality educational programming to homes and classrooms throughout Latin America. Furthermore, it has helped develop AME (Actualización de Maestros en Educación), an innovative distance learning program that provides training to schoolteachers across the region via satellite television. [172]

Some of the groundwork for distance education in that part of the world already has been laid. Costa Rica and Venezuela, for example, have distance education institutions: Universidad Estatal a Distancia in Costa Rica and Universidad Nacional Abierta in Venezuela. In addition, the Monterrey Institute of Technology (ITESM) in Mexico has been in operation since 1989. Its coverage extends to various Latin American countries. It has become a pioneering institution in distance education in the Americas. Using learning networks and advanced technologies, the Virtual University provides graduate academic programs, continuing education programs for companies, and programs for elementary and secondary school teachers.

Finally, CREAD is an Inter-American, non-profit distance education consortium. Founded in 1990 at the International Council for Distance Education World Conference in Caracas, Venezuela, CREAD's has carried out a number of important missions in its nearly 20-years of operation including hosting a number of important conferences.

LIVENING UP K–12 EDUCATION

With apologies to all the world's hard-working teachers, for kindergarten through 12th grade, online education is beginning to put new life into curriculums.

Consider the Global Schoolhouse (www.globalschoolnet.org). Begun in 1992 with a grant from the National Science Foundation, it is an internationally recognized K–12 project funded through a public/private partnership that includes the National Science Foundation, Cisco Foundation and Microsoft.

Students throughout the U.S. and globally share information on cooperative projects. They use email, the World Wide Web, and live video teleconferencing with their PCs. Projects they've worked on include alternative energy sources, solid-waste management, space exploration, and weather and natural disasters. [173]

The Global Schoolhouse is the brainchild of Global School Net Foundation, a nonprofit corporation, which wants to be known as a major contributor to the "philosophy, design, culture, and content of educational networking on the Internet and in the classroom," according to the Global Schoolhouse home page. From its beginnings in the mid-1980s when it was founded by schoolteachers in San Diego, California, with no budget and minimal support, Global School Net has emerged as an internationally recognized piece of the global education infrastructure.

Then there's KIDLINK (www.kidlink.org), founded by Odd de Presno, a Norwegian journalist and author of computer books. KIDLINK is sort of a global kids' coffee klatch, where kids from classrooms all over the world can discuss topics, through the KIDFORUM. [174]

In Florida, Space Coast Middle School in Port St. John is completely wired. The school opened in August 1995 with 1,650 sixth-, seventh-, and eighth-graders. It was designed as a model technology school for the Brevard County, Florida, school district.

In Maryland, the Montgomery County public schools instituted a Global Access plan to electronically connect classrooms, media centers, and offices so students and staff can access information and communicate locally through an "Intranet," computers with software tools and access to networked resources in each school's media center. The Global Access plan later became the Technology Modernization Program (Tech Mod), which is designed to upgrade over 40,000 computers over a four year period, provide schools with upgraded networks, a lower student to computer ratio, and on site staff development in curriculum integration. [175]

At Hickory High School in Hermitage, Pennsylvania, a library I funded is one of the most technologically advanced in the country. It was designed from the ground up to be a model for high school libraries for the 21st Century. It contains a distance learning room, where students and instructors can interact in real-time from different locations, a space bridge lined with cyber modules and a space with an information wall at the end of the bridge.

The U.S. cable industry has been wiring schools since 1989. Cable companies also are creating wide-area fiber-optic networks connecting community educational resources in rural and urban U.S. locations. For example, in Mercer County, New Jersey, Comcast Corp. and a 14-member local educational consortium created MercerNet. This network links all Mercer County school districts, Mercer Community College, and a local science center with one another and with each of the county's public libraries and state colleges.

The network provides interactive TV for distance learning and community programs, high-speed cable access to the Internet, and high-speed data connectivity via cable, interfaced with multimedia video libraries in and out of the U.S.. With help from a $700,000 grant from the National Telecommunications and Information Administration, 14 interactive video classrooms with multiple-data channels were connected to MercerNet.

Finally, there's the Jones e-global library®. This library is a powerful full-service online resource management portal (www.egloballibrary. com). It's fully customizable, available 24-hour-a-day, is geared to serve the needs of time-deprived adult learners, includes over 31,000 electronic books, connects to thousands of databases and has a confederated search engine.

ONLINE LEARNING IN CANADA

Canada, long a leader in distance learning, has continued moving forward. Ontario, for instance, has three consortiums: the Ontario Strategic Alliance for eLearning, POOL and the French Language Boards which share resources across multiple schools. Across Ontario, there are an estimated 25,000 students taking online courses, according to iNACOL. In Quebec, 90 percent of teachers use computers for their administrative work but close to 50 percent are still not using them for their teaching and learning activities indicating that teacher training is still a major issue. [176]

THE EXPLOSION OF INTEREST INTERNATIONALLY IN WEB 2.0

The growth of Web 2.0 and the interest in the Web is truly an international phenomenon. Consider these statistics from 2008 alone from comScore, a leader in measuring the digital world:

- 85 percent of Internet users in Brazil visited a social networking site in September
- More than 14 million U.K. Internet users visited a blog in August
- Over 27 million people watched more than 3 billion videos online in the U.K. in June
- YouTube drew more than 5 billion views in the U.S in July
- 25 million people watched 2.3 billion videos in France in May and
- MySpace.com attracted 1.2 million visitors in Japan in June. [177]

THE QUESTION OF ACCESS

The effort to get primary and secondary school classrooms in the U.S. wired is commendable. But are enough potential distance learners able to use the Internet or other online systems, not just in North America, but worldwide?

There's good news and bad news on this subject. The statistics on online usage are impressive. In 1985, for example, there were only 300,000 registered email users worldwide. In 1993, there were 12 million people who used email and other online services just in the U.S. [178] By 2008, the number of email users and the number of email messages sent were huge and looked to grow even larger. The Radacati Group predicts that the number of worldwide email users will increase from 1.3 billion in 2008 to almost 1.8 billion by 2012. More than one in every five persons on the earth use email. Worldwide email traffic totaled 210 billion messages per day in 2008, and by 2012, this figure will nearly double to 419 billion messages per day. Sixty three percent of all traffic was consumer in 2008, and 37 percent were corporate email messages.

In 1976, only 50,000 computers existed in the entire world. In early 1996, more than 50,000 computers were sold worldwide every 10-hours. [180] In 2008, the number of PC's in use reached over one billion, according to *Computer Industry Almanac*. [181]

In 2000, the U.S. and Hong Kong had the highest computer-per-person ratios in the world at one for approximately every two people. Canada's ratio was not far behind, at one for every four Canadians, and Australia's was slightly higher. The U.K. also boasted one computer for every four of its citizens. [182] By the end of 2008, the U.S. had 274 million computers-in-use while the world total surpassed 1.23 billion units in 2008. [183]

In 2000, there were 738 million mobile cellular subscribers worldwide. By 2007, that number grew to over 3.3 billion. There were 390 million

Internet users in 2000; by 2007, that number grew to nearly 1.4 billion Internet users, according to the International Telecommunications Union. [184]

Among the developed nations, and in some of the fast-progressing lesser-developed countries such as China and India, both computer ownership and Internet connectivity are increasing at a pace that has outstripped even the most optimistic forecasts of the mid-1990s. In late 1995 and early 1996, Internet connectivity for schools in the U.S. still seemed a distant goal. At the end of 1995, only 35 percent of U.S. public schools had access to the Internet; only three percent had any individual classrooms connected. [185] But a dramatic change occurred over the next decade.

The nation's economic boom and rapid deployment of technology gravitated into the public education sector more quickly than most experts assumed it could. In a feat of technology adaptation and market penetration whose only likely parallel is the production and deployment of military arms in the U.S. during World War II, most of the nation's public schools were wired in a little over three years. The percentage of instructional rooms with Internet access increased from 51 percent in 1998 to 94 percent in 2005. And nearly all schools had Internet access by 2005, and the average public school contained 154 instructional computers by then. [186]

Much of this progress was made possible by the cable industry and by grants and assistance from U.S. companies in the computer and technology industries. A large part of the task was also funded by public and government investment channeled through normal appropriations.

Still, the schools that did not have Internet access were a concern, giving rise to claims of a "digital divide" between those with adequate computer and online access and those without it. Those without were primarily poor inner-city school districts and some of their rural counterparts.

A 1996 article in *Newsweek* magazine illustrated this circumstance all too clearly. The story was about a California-based high-tech chipmaker, MIPS, that wanted to help rural and inner-city schools gain access to the Internet. MIPS gave more than $55,000 to a school in Brooklyn, New York. The school didn't use it to buy computers. "They had a more urgent need," the article reported. The school administrators used the money to buy desks and chairs. [187]

The story pointed out some glaring inequities in U.S. schools, but make no mistake—the trend toward using the latest telecommunications technology in teaching either on site or at a distance and at every level is no passing fad. Cost of technology is no longer as prohibitive, at least in the developed world, though lack of teacher and administrator computer literacy is a continuing concern.

INTERNATIONAL CONNECTIVITY RAMPS UP

A disturbing statistic is often tossed into conversations about technology by telecommunications analysts: only half the world's population has ever made a telephone call.

In its 1999 Human Development Report, the United Nations devoted a chapter to "new technologies and the global race for knowledge." [188] The primary point of the authors is that member countries of the OECD must focus on making technology access equitable at all levels of society. But the report's second level of impact is more graphic. Many of the trends and comparisons cited in the report give startling credence to the breakneck character of the globe's rush toward adaptation and convergence of computer, Internet, and other telecommunications technologies and other scientific breakthroughs. Any way you choose to look at it, nothing quite like it has happened before.

For example, the report noted:

- It took 38 years for radio to achieve 50 million users worldwide; it took the World Wide Web four years;

- Software exports from India rose in value from zero in 1980 to $1.5 billion U.S. in 1997;

- English is used in almost 80 percent of Web sites, including the graphics and instructions, even though fewer than one in ten people worldwide speak English;

- Industrial countries hold 97 percent of all patents worldwide.

- Use of intellectual property rights is alien to many developing countries; and

- Building telecommunications infrastructure does not ensure people will have the skills to use it.

That last point brings us back to the focus of this book, education and new ways to deliver it. The same United Nations report outlines a growing trend to provide the technology, training, and educational services necessary for people to convert their learning into employable skills.

In terms of grassroots initiatives toward cyber education worldwide, the report cites several examples: Building people's capacity to use the Internet starts in schools. The government of Costa Rico had installed computers in rural schools across the country to give all pupils a chance to learn the new skills. In Hungary the ambitious Sulinet (Schoolnet) had enabled students in more than two-thirds of secondary schools to browse the Net from their classrooms. [189]

Still, there is much to be done. Despite the widespread, rapid deployment of computers and the Internet, at the dawn of the new millennium, less than three percent of the world's population was connected to the Net. [190]

The billions of mobile phones, however, are increasing Internet access worldwide at a scale never seen before. The confluence of the mobile phone and the World Wide Web will increase Internet penetration exponentially. It will occasion a tsunami of opportunity and innovation.

WHAT'S NEXT?

Is there something out there in the ether that will revolutionize distance learning beyond the new horizons of the Internet and the World Wide Web? Perhaps something simpler, faster, cheaper than laying phone lines and buying desktop PCs will make even faster growth possible, upping the connectivity to 20, 30, or even 50 percent and beyond.

For starters, there are increasingly affordable personal computing/Internet access terminals; some are even available for free in exchange for the user signing up for several years of Internet access. Satellite delivery of Internet services can also speed up the accessibility of Internet service providers (ISPs) in rural areas and developing countries. The proliferation of cellular phones in such countries as India, China, Mongolia, Russia, and various archipelagoes such as Indonesia has accelerated affordable telecommunications access in these countries by at least a generation ahead of most predictions in the early 1990s.

One trend that has driven increased access has been the deregulation of telecommunications markets. In country after country, from Germany and the U.K. to Singapore and Hong Kong, as telecommunications access has been opened to private, market-based ownership, the price for consumer service has fallen. Now it looks as though access to computing power may enjoy a similar leap forward.

Greater bandwidth is coming as quickly as companies can install it, undersea cables can be laid, and satellites can be launched. The possibility exists that major utility companies, known traditionally for providing power to communities, might become telecommunications and cable TV competitors — by stringing fiber-optic lines through natural gas pipes, laying new cables along utility rights-of-way, and sending telecommunications voice and data over high-power transmission lines.

The gigapop. Another link in the greater bandwidth revolution is the gigapop. A pop is a point of presence, or the connection of a network to the Internet. A gigapop is a supersized pop, which allows information to move at a rate of 1,000 megabits per second, some 100 times faster than the speed of a typical university computer network and 4,000 times faster than the speed of a typical household modem. Gigapops are already being employed on campus systems and are greatly speeding the delivery of information — and classroom options — from professors to computer-equipped lecture halls.

A student who is at home and who has the option of switching to a cable modem or other high-speed connection can also benefit from the wider, faster bandwidth. The Seattle Community College installed a gigapop in 1999 at a cost of $10,000, plus contributions of reduced service prices from local TV stations and a partnership with the University of Washington. [191]

Internet2. Already operational, Internet2 is a high-performance network operated by a consortium of over 200 universities, 70 corporations and 45 affiliate members. Established in 1996 to build a better, faster Net that would be their own, this is a very different Net from the consumer-access Internet.

Internet2 uses a hybrid mix of university, government, National Science Foundation (NSF), and other existing communication networks plus an array of new network tools not commonly used in the primary Internet — sort of a son of Superman.

Internet2 provides a high performance networking infrastructure connecting nearly 300 member organizations and more than 42,000 research and education institutions such as high schools, museums and libraries across the U.S.. Rather than replacing the Internet, Internet2 aims to provide capabilities 3 to 5 years ahead of the commercial Internet and a test bed for future technologies that can be integrated into the Internet in the future. [192] It also addresses the speed of response and alleviates connection interruptions, a chronic problem of the current Internet. The universities do intend to develop and prove new technologies that will make an eventual super Internet for everyone much superior to what is available today.

International alliances between leading universities utilized Internet2. The Singapore-MIT alliance, for instance, started in 1998. This alliance consisted of the Massachusetts Institute of Technology, the National University of Singapore and Nanyang Technological University. The alliance allowed for synchronized learning using the Internet2 infrastructure. [193] In 2009, courses in the alliance used a live lecture format delivered via videoconferencing with supplemental data content provided over an application sharing link.

Video streaming technology. This technology has been available since the mid-1990s, and in 2000 college faculty and K–12 teachers were scrambling to acquire the skills to use it, just as they had to develop Web skills in the last four years of the 20th century. For some, the adaptation of video streaming will intensify the debate over the "edutainment" of education; for others, it will be seen as a tool that put both instructors and students into virtual environments that previously could only be described, read about, or viewed secondhand.

How about "smart card" technology? A smart card looks like your garden-variety credit card, but it's capable of tracking and organizing a distance education student's learning programs with the same degree of efficiency and pervasiveness as credit cards track individual finances. [194]

As the personal communication system (PCS) replaces the telephone, allowing everyone in the world to have the same phone number for his or her entire life, smart cards will become the way individuals access and interact with all manner of information, experts predict. For example, let's say you are on vacation in Germany and need to check up on an assignment for which you've forgotten the due date. Attached to your PCS is a smart card with a microelectronic circuit that keeps track of, among other things, your educational progress. Slip it into a smart card "reader" or perhaps a computer terminal, and it will display the current status of all your coursework.

According to Barry Barlow, an educational technology expert, smart card technology will be interactive. Most of the work on smart card technology has been and is being done in Europe.

What about virtual reality? Virtual reality is "a complete environment" assembled and managed by a computer software program. Instead of manipulating two-dimensional images while sitting at a keyboard, the virtual reality participant dons a special interface — right now a pair of goggles — that allows him or her to become part of a three-dimensional program. [195] For example, ScienceSpace, an evolving "suite of virtual worlds" that helps students master difficult science concepts, was created by Chris Dede, now the Timothy E. Wirth Professor in Learning Technologies at the Harvard Graduate School of Education , and by colleagues. [196]

Dede wrote that virtual reality is more like diving into an aquarium as opposed to looking at fish through the aquarium's window. Three virtual worlds have been developed at George Mason:

- NewtonWorld allows students to dive into the world of one-dimensional motion;
- MaxwellWorld puts students inside electrostatics; and
- PaulingWorld allows students to experience molecular structure from the inside out.

By becoming part of the phenomenon, virtual learners gain "direct experiential intuitions about how the natural world works," Dede contended. [197]

Perhaps the most enabling of recent technological advances and the globalization of education described in this chapter is mobile, wireless technology. Accelerating the adoption of mobile wireless will be the development and deployment of mobile applications for smartphones, the increased purchases of NetBooks, the increasing power of 3G networks, the falling prices of handsets and the incessant drive for sizable profits from a relatively competitive environment.

While the potential for mobile wireless for educators currently is largely uncharted territory, its potential is immense. For this reason, this edition of Cyberschools includes the following chapter which focuses on this enormous opportunity.

CHAPTER SIX
The Onrushing Revolution in Mobile Learning

*"The sheer ubiquity of mobile phones amounts
to 'the biggest leap in history, bigger than
the printing press, which, after all, stayed in the
hands of very few people'"*

Katrin Verclas quoted in The Economist [198]

One of the most startling technology trends since the last update of *Cyberschools*, and indeed one of the most important trends so far in this century, is the tremendous growth in wireless subscribers and the huge increase in the number of owners of mobile phones.

Consider these stunning statistics:

- By the end of 2008, there were more than four billion mobile cellular subscriptions worldwide [199;]

- The number of mobile customers from 2002 to 2007 in the U.S. alone almost doubled and currently there are over 255 million customers, as Table 6 indicates; and

- By the end of 2008, there were over three times more mobile cellular subscriptions than fixed telephone lines. [200]

In many countries, the number of cell phones exceeds the number of people. [201] The explosion in mobile customers highlights another important trend: globalization and, in particular, the continued ascendancy of India and China as major global economic powerhouses whose growth generally far exceeds that of the U.S. In 2007, China had over 500 million mobile customers, far outpacing the U.S. as shown in Table 6. India has moved from a small number in 2002 to over 200 million customers in 2007, an astonishing increase in just five years.

Table 6. Growth in Mobile Customers [202]

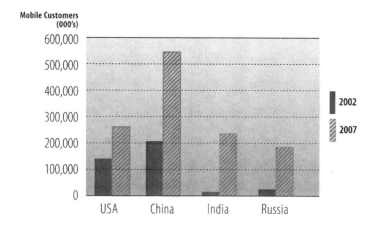

This eye-popping growth is leading organizations throughout the world to explore how to tap into the learning potential of the "third screen" available in the wireless device; the first and second screens are, of course, the television and the computer.

In developing countries, the impact of the mobile phone is significant. The Hewlett Packard Foundation reported that:

> "Mobile phones are already the dominant user platform in developing countries; it is expected there will be 2.5 billion users in those regions by 2010. By that time, if current trends continue, the typical mobile phone will have the processing power of today's desktop PC... The bottom line is that, for the vast majority of people in developing countries, their 'PC' and Internet access device will be a mobile phone, a handheld computer, or a hybrid of these devices." [203]

Even the term "mobile phone," as of 2009 has become a misnomer. Don Tapscott suggests calling them a "buddy or even a digital copilot" as these devices have become "small, powerful computers that are part voice communication, part BlackBerry, part iPod, part Web browser, part texting device, part digital camera, part video camera, part voice recorder and part GPS compass." [204]

Mobile phones and consumer electronic devices are morphing into something like a computer and computers are embracing features of mobile phones. Device innovation is accelerating. As this happens, entrepreneurs and entities inside and outside of the traditional education and training communities will initiate changes into those ecosystems.

MOBILE PHONES ARE "LENSES ON THE ONLINE WORLD"

In April, 2008 *The Economist* reported that a staggering 3.3 billion-plus people subscribe to a mobile phone service. This is nearly triple the number that existed only in 2002. [205]

The explosion of mobile usage will allow Internet usage to expand even more dramatically. One survey of cell phone users in the U.S. Britain, France, Germany, Italy and Spain found that more than a third of those who had Web mobile access used it.

Globally, there are more cell phones now than PCs. "I always hear about the cell phone being the third screen, but I think about it as the first one. It's with me all the time," Bob Greenberg, CEO of R/GA, an agency based in New York, told the *New York Times*.[206]

The potential for new global educational applications via mobile phones or digital devices is the new frontier, particularly in developing countries. An excellent report for the Hewlett Foundation concluded that "The bottom line is that, for the vast majority of people in developing countries, their 'PC' and Internet access device will be a mobile phone, a handheld computer, or a hybrid of these devices."[207]

Four billion people exist at the base of the economic pyramid (BOP), defined as those people who live in relative poverty with incomes below $3,000 in local purchasing power. A report by the World Resources Institute and the International Finance Corporation asked, "Will phones become the Internet platform for BOP households and rural communities?" and concludes, "Several factors suggest that they will."[208]

> "Mobile phones already have an enormous lead over computers in developing countries... Phones are relatively easy to master, generally require no sophisticated technical support, and, as voice-based devices, pose no literacy barrier. Phones are less expensive than computers—basic GSM models designed for developing countries are approaching US$30—and service is often offered through prepaid business models that are more affordable for BOP consumers," according to the report.[209]

The confluence of the World Wide Web and mobile phones, indeed, may be *the* trend to watch in the decade ahead. More than three quarters of the respondents to a survey conducted by The Pew Internet & American Life Project agreed that in 2020, the mobile phone—now with significant computing power—will be the primary Internet connection and will be the *only* one for a majority of the people across the world. "At this point, the bottom three quarters of the world's population account for at least 50 percent of all people with Internet access — up from 30 percent in 2005." [210]

The growth in mobile phone sales is as fast as lightening. It took 20 years for the first billion mobile phones to sell, four years for the second billion and just two years for the third billion. [211] In the Pew report, Susan Crawford, the founder of OneWebDay, wrote that by 2020, billions of people will have joined the Internet who do not speak English. "They won't think of these things as 'phones' either — these devices will be simply lenses on the online world." [212]

As things currently stand, the cell phone business model may work better than the Internet model. Subscribers have been trained to pay for mobile phone products whereas people have been trained to pay nothing via the Internet. The existing habit patterns could likely continue, so economic models that wouldn't work on the Internet could actually work on the mobile platform. As the cell phone fuses with the Internet, therefore, an abundance of opportunity exists.

THE M-LEARNING BOOM

In 2006, over a billion mobile phones were built and shipped. The implications are large, indeed.

> "The combination of social networking and mobility lets students and colleagues collaborate from anywhere they happen to be. Add to that connectivity the multimedia capacities of phones, and the storage they offer for podcasts, videos, photos, PDF files and even documents and spreadsheets, and it is not hard to see why phones are increasingly the portable tool of choice," according to the *Horizons Report.* [213]

The new technology is out there. What we do not have now, we will eventually create anew. It's all possible. What makes sense for distance educators to use remains to be determined. It is hoped the technology that can provide the greatest access to education for the greatest number will be at the top of distance educators' shopping lists worldwide.

Since 2002 when the last edition of *Cyberschools* was published, considerable advancement has been made in the field of mobile learning. In 2002, a conference series called *mLearn* began, and in their conference of 2008 a brief background was given about the growth in mobile learning from 2002 to 2008:

> "In the intervening period, mobile learning has matured and consolidated," one expert wrote. "It now has a newly created peer-reviewed academic journal, the *International Journal of Mobile and Blended Learning* and a professional body, the International Association for Mobile Learning. It also has a vibrant online community, much of it logging onto the Handheld Learning forum (http://www. handheldlearning.co.uk/) and a critical mass of prestigious international conferences." [214]

It has gained increasing acceptance. Marc Prensky, CEO of Games2train and author of *Don't Bother Me Mom – I'm Learning* and *Digital Game-Based Learning*, said, "Mobile learning is going to happen much faster than people think [and] it's going to come very quickly to schools around the world. Students and teachers need to partner with each other because they can't do all the things they used to do." [215]

CORPORATE M-LEARNING CASE STUDY

Many groups are engaging in, and experimenting with, mobile learning such as K–12 schools, universities, continuing education providers and many other institutions. One corporate example is particularly worth noting.

Merrill Lynch. Merrill Lynch, which was purchased by J.P. Morgan in 2008 and despite its myriad problems, has shown some highly creative ways to train its traveling employees. Merrill Lynch, in fact, implemented with its GoLearn platform the next generation of corporate distance education: mobile learning. While Merrill Lynch found its live, instructor-led training valued, back in 2006, getting 60,000 people from three different business units in 38 countries in the classroom was difficult, to say the least, according to *Chief Learning Officer* magazine. [216]

A new training solution existed that married technology with learning. Merrill Lynch had more than 21,000 BlackBerry devices in use globally with 500 new devices being added monthly; and most of its investment bankers were constantly on the road. So in 2007, Merrill began training via the Blackberry through three compliance training programs back in 2006. The results from Merrill's GoLearn project were remarkable. Participants took their courses in 45 percent less time saving about four to six hours in time per year; the completion rate was 12 percent higher at the 45-day milestone. Employees have also scored higher on the competency exam than colleagues studying in the traditional online formats. [217]

Clearly, the confluence of the Internet and mobile technology has great promise for education. It doesn't stop there. Combine all of this with evolving, always-on, broadband-enabled, easily accessible next-generation Internet "cloud computing" concepts and you have something immensely appealing. This empowers people to more effortlessly move back and forth between the tangible world and the intangible (virtual) world. In many cases, we will not even be aware we are doing it. Instead of dealing with many pieces of software and hardware, things are simplified and one can just plug into the "cloud" and everything happens without all the mental strain encountered previously.

The early on point of focus for much of this will probably be the cell phone (or its competitors) which is morphing into a digital device that delivers an increasing cornucopia of services. It constitutes the largest, by far, distribution system in the world and is generating huge, ever increasing revenues as it adds new services.

In my judgment, it is this increasing revenue stream that will provide many of the billions of dollars required to develop and pay for the technology that is needed to evolve new services that will define the future. It is difficult to imagine all the ways that this complex mosaic will further impact education. However, it appears obvious that it is capable of redesigning both the platforms upon which we educate and the processes with which we do it. The digital onslaught and the change embedded in it are now looking for education and there is no place to hide. As we move forward, expect the unexpected.

CHAPTER SEVEN
Distance Learning: Defining the Market

"For a technological advance to be truly basic, it must change the entire world. In the last half century, the coming of television, of jet planes and of solid-state electronics has presented examples of world-changing technological advances. Dwarfing these, however, is the communications revolution."

Isaac Asimov

Where there is a need, there usually is a market. Education is no exception. The global market for distance education delivery is burgeoning.

Multiple statistics demonstrate that this growth is not just a U.S. phenomenon but a global one. Online enrollments continue to rise at rates far exceeding those of the total higher education student population. According to findings from the Babson Survey Research Group and Sloan-C:

- Over 3.9 million students were taking at least one online course during the Fall of 2007; this is a 12 percent increase over the number reported the previous year.

- The percent growth rate for online enrollments was nearly 13 percent by the Fall of 2007 while the growth of the overall higher education student population stands at 1.2 percent.

- Over twenty percent of all U.S. higher education students were taking at least one online course in the fall of 2007. [218]

The total U.S. online higher education market stood at $8.1 billion in 2006, according to Eduventures, a Boston-based research group. [219] And online tuition revenue totaled $7.1 billion in 2005, up from $2.4 billion in 2002. [220] Eduventures estimates that in 2007 there were 1.5 million 100 percent online students representing eight percent of all students at U.S. degree-granting Title IV schools and 20 percent of adult students. [221]

When we look at the developing world's need for education, we see that these estimates are just the tip of the iceberg. The population is exploding. As of 2009, there were nearly 6.8 billion people on the globe. That number is expected to grow to 9 billion by 2040. Keep in mind that during just a 40-year period (1959 to 1999), the world's population doubled from 3 billion people to 6 billion, according to the Census Bureau.

Total U.S. enrollment in degree-granting institutions increased in a 10 year period (1995 to 2005) from 14.3 million to 17.5 million, an increase of 23 percent. Much of the growth was in female enrollment; the number of females grew by 27 percent and the number of males rose by 18 percent. [222]

Perhaps this doesn't sound too daunting, until we are confronted with the fact that there were only 15 million postsecondary classroom seats in the U.S. in 2000. Paradoxically, institutions were having great difficulty funding building programs and recruiting faculty to serve students who often were only able to attend part-time. Add to the projected student body another 90 million U.S. adults who wanted part-time continuing education beyond the traditional two- and four-year programs, and we see a considerable bottleneck. This, in addition to the sudden availability of affordable technology, may explain why traditional U.S. institutions developed more distance learning capabilities between 1995 and 2000 than they did during the entire previous half-century.

CHINA'S EDUCATION MARKET

Other solutions are being applied or devised in other countries, depending on the technology available. China's higher education infra-structure illustrates my point. By 2004 in China, the total enrollment of tertiary education reached more than 20 million.[223] The higher education system in China counts about 2,000 universities, colleges, and adult education institutions that enrolled over nine million students in 2002, according to the China Ministry of Education. Over three million of these were admitted into higher education institutions.

As a result of the Modern Distance Education Project (MDEP) initiative, 68 e-colleges were established between 1998 and 2005. By 2003, there were 2.3 million enrollments registered for distance education programs in the 68 pilot universities in China. Of those, 57.8 percent were registered for undergraduate courses, 41.7 percent for vocational courses and .5 percent for postgraduate courses.[224]

Internet usage is growing in China but is still nascent. Some 94 million people in China became Internet users by the end of 2004, representing a yearly increase of over 18 percent since 1997.[225] "The number of online learners is growing," according to a report by iNACOL, "but the increasing number of e-Learners still accounts for a very small percentage of China's population. Students under the age of 22-years account for 23 percent of online learners; ages 22 to 29 account for 50 percent; and 27 percent of 30-years and older aged students are participating in online learning. The majority of these students are from the urban areas of China where students have access to the Internet through broadband connections."[226]

The size of the market in China for online education has grown fast. In 1999, the online educational market amounted to $1.7 billion US dollars; by 2004, it rose to $23.1 billion US dollars, a very high growth rate of 66 percent.[227] A devout believer in distance education, China has the China Central Radio and Television University, with more than 2.3 million students. It has 52,600 full time academic staff.[228]

In the last decade, traditional media have been challenged in China by the increase in Web-based course offerings. Chinese distance education has evolved through three stages from "correspondence-based education; [to] broadcasting/television-based education since the 1980s and advanced distance learning based on information and Internet technologies since the 1990s," according to an article in the *International Review of Research in Open and Distance Learning.*[229]

There are challenges, to be sure, in delivering electronically based higher education to different parts of the world. In Japan and Korea, delivering higher education on the Internet is not a problem, although overcoming the language barrier presented by English-based education is a hurdle. In India, where expertise and technological savvy are in place and where English is either the first or second language for many people, the caste system creates barriers for many potential distance learners.

The issue for Asia—with the exception of India and a few other countries—is that the culture of learning is different. It is, by and large, one way, from teacher to student, and is based more on memorization than is U.S. style higher education. It is necessary to keep both Western and Asian cultures.

In the rest of English-speaking East Asia—Singapore, Malaysia, parts of Indonesia, Hong Kong, and the Philippines—there is an enormous demand for higher education. The telecommunications infrastructure has been developed quickly, making distance education feasible in the metropolitan areas. Australia and New Zealand are long-time leaders in distance education. The market in both countries is both wired and proven.

EUROPE'S DEMAND AND CHALLENGES

There is demand in Europe, although the potential student population is smaller. British Open University provides the potential for higher education at a distance to nontraditional students from all over the European Union. In Europe, the challenges facing distance education are cultural, not technical. Especially in western Europe, computers are increasingly part of the furnishings in the homes of most of the intelligentsia and middle class. Satellite technology is advanced all over Europe. As in the U.S., TV is everywhere. But Europe in general does not have a tradition of promoting open access to education, despite the great success of British Open University.

The market also is substantial in South America and Africa. These populations are large and growing. In countries where there is no telecommunications or cable infrastructure, it is cheaper altogether to bypass these traditional infrastructures and move directly to mobile. Cyberschools will be able to use mobile as a platform.

In Africa, the challenges are political, social, and economic. In the mostly Muslim north, there is a small group that can afford to tap into technology-delivered distance learning. But, in general, the masses of people in that region don't have adequate access.

In central Africa, the same factors are at work, but to an even greater extent. And the percentage of the population that can afford more than the most basic standard of living is minuscule. In South Africa, the situation is better. It's interesting to note, however, that South Africa has had television only since the mid-1980s.

For generations, South Africa's black and colored population was provided with less than adequate education under the policies of apartheid. Now, the South African government hopes to remedy the public education deficit using distance education technology. But first things first. Many rural areas and townships have yet to receive electricity.

When South Africa celebrated its 10th year of democracy in 2004, the black population represented over 80 percent of the population yet only represented 53 percent of students enrolled in South African universities. Over 65 institutions offered distance learning in higher education in South Africa. The University of South Africa (UNISA), the nation's leading university and an open learning and distance education institution, had 265,794 registered students in 2007, up over 15 percent from 2006. [230]

In North America, the U.S. and Canada are wired, and Mexico's Monterrey Institute for Technology and Education (MITE) is one of the organizations leading the way with satellite-delivered and Internet-augmented distance education. Two of the Institute's cornerstone projects are the National Repository of Online Courses and the MacArthur Series on Digital Media and Learning.

Mexican professors also have their own listserv on the Internet, called Profmex, a consortium for research about Mexico is made up of some 80 institutional members and 500 individual members. Measured against countries in the developing world, Mexico's population is highly educated; more than 90 percent of its citizens are literate. Spending per student increased by 49 percent for primary and secondary students and by 67 percent for tertiary students between 1995 and 2003, according to the OECD. And Mexico's share of public spending invested in education is the highest among OECD countries and almost twice as high as the OECD average. [231]

In Central America and South America, politics, culture, and technology access are the challenges. Central America, which includes the Caribbean, has a significant English-speaking population, but politics are turbulent, and universal access to technology is a distant goal. In South America, with the exception of Chile, Brazil and perhaps Argentina, the same challenges apply, and the languages are Spanish and Portuguese. In addition, faculty unions in universities are entrenched and powerful. Although many faculty members are exploring electronic platform delivery, many others do not welcome technology-based distance education.

Brazil is the largest country in South America and the fifth largest in the world with 190 million people. Distance education has been slow to progress for a number of reasons; the government, for one, heavily regulates the educational sector, often by decree. Furthermore, about 20 percent of the population currently live under the poverty line and are functionally illiterate. Despite these obstacles, there are currently three million learners enrolled in courses offered at a distance throughout Brazil. Of these three million distance students, 1.5 million are registered in corporate initiatives. [232] Under the Ministry of Education, the Open University of Brazil (UAB) began in 2005 and is a consortium of several state universities, offers a hybrid model of distance education. Also, in 2007, a similar initiative started at the professional education level. The Ministry of Education inaugurated the eTec Brasil — Open Technical School System of Brazil. Corporations such as Ambev in beverages and Accor in hotels are investing in continuing education for their executives. [233]

Despite the many hurdles to distance education, however, I believe information highway technology will be available in most parts of this vast planet sooner than most people think.

For one thing, costs are falling dramatically with technology. In March 1996, U.S. computer maker Gateway introduced Destination, a combination TV-computer for the home market. [234] The new machine used a Pentium chip–based personal computer with special accessory cards for high-quality sound and video and was attached to a 31-inch computer monitor. It included a wireless keyboard and a remote control. At the time Destination was launched, its suggested retail price was $3,800. Since then, WebTV, which was quickly purchased by Microsoft, entered the market with a set-top box and keyboard that provide the basic hardware needed to connect most TVs to the Internet for $200. This was only the first of what became a flood of convergent technology and big-screen home computer/video systems that will effectively place classrooms, laboratories, and professors' offices in the living rooms of the world. [235]

Richard Green, President & CEO of CableLabs noted, "The capacity to distribute video world-wide is now available. Video has the extraordinary power to educate and inspire learning. The technical capability to reach around the world with educational offerings transmitted via television is now a reality only dreamed of in earlier times."

A CAMPUS FOR EVERY CONTINENT

In 1966 the most advanced transatlantic telephone cable could carry only 138 simultaneous conversations between Europe and North America. In 1988, the first fiber-optic cable carried 40,000 simultaneous conversations. The fiber-optic cables of the 1990s carried nearly 1.5 million. [236] In 2000, with satellite technology it was possible to beam distance learning programs to every continent of the world. A global electronic campus using existing telecommunications satellites and fiber-optic cable is possible. The campus's reach would encompass the satellites' footprints on Earth, and much of the content could be delivered by the Global Internet.

The infrastructure to pick up satellite signals is growing rapidly. In 2008, industry experts estimated that almost 110 million households in India will have broadband services over the cable TV network. Television will be the bridge for the digital divide. With very little additional cost, broadband facilities can be provided to over 50 million cable TV households right away, Vijay Yadav, a Managing Director of IPTV service provider UTStarcom, told reporters at the fifth Broadband Tech India conference, according to *India eNews*. "If this becomes a reality, then we can witness broadband subscriptions to go far beyond the target of 20 million by 2010," he added. [237] With the availability of cheap computers, India in particular is rapidly becoming a global center for computing services. It has the opportunity to be a leader in distance education applications.

WORLDWIDE ELECTRONIC CORPORATE EDUCATION AND TRAINING

Higher education delivered electronically worldwide is not just constrained to the kind that culminates in a university degree. It also is appropriate for executive education and training. In that realm, the educational delivery pendulum is swinging toward technology at an ever-increasing rate. Corporations, now lean and mean, can't afford the luxury of sending managers to one- or two-week training programs off site. Indeed, *Business Week* magazine quoted one U.S. CEO as saying, "If we can do without someone for a week, we can probably do without them for good." [238]

Faced with retraining over 50 million American workers, corporate America is using distance learning, both internally and externally, for all aspects of training, the U.S. Distance Learning Association stated. Using distance learning to train employees more effectively than with conventional methods saves corporations millions of dollars each year.

The U.S. corporate e-learning market reached $9.7 billion in 2007, according to the IDC. The worldwide market reached $15.9 billion. While some of the growth is attributable to exchange rate advantages in Europe, the IDC added, "most of the growth represents real increase in demand." [239] U.S. organizations overall spent over $134 billion on employee learning and development in 2007, according to the American Society for Training and Development.

WHO'S DELIVERING DISTANCE CORPORATE EDUCATION

Universities, both for profit and nonprofit organizations, and public/ private partnerships are beginning to beam corporate education and training almost everywhere. The delivery modes vary from real-time, video-based instruction to Internet-based virtual training programs.[240]

Founded in 1984, National Technological University (NTU) was the first accredited distance-learning engineering university in the nation. Hewlett-Packard, IBM, Lockheed Martin, and Motorola supported NTU early on as a means to keep their engineers on the cutting edge of technology. In 2005, NTU merged with Walden University to become the NTU School of Engineering and Applied Science. Their programs offer world class scholars from universities such as MIT and Purdue. Besides Hewlett Packard, Lockheed Martin and Motorola, the school has worked with Boeing and Raytheon.[241] Walden University is the flagship of the Laureate International Universities network, a global network of more than 42 accredited online and campus-based universities in 20 countries. They serve almost 500,000 students on over 100 campuses world-wide.

In early 1999, Jones International University (JIU), founded in 1993 and headquartered in Colorado, became the first totally online university to achieve accreditation from one of the U.S.'s regional accreditors.[242] JIU's goal was to attain the same level of accreditation required of established site-based universities, prove that it could be done and thereby establish a required high level of quality for the then nascent online education environment. This was a pre-emptive step to prevent the acceptance of a lower level quality standard for online education which could have diminished its potential as well as its ability to democratize education.

Will and Ariel Durant said, "If equality of educational opportunity can be established, democracy will be real and justified. For this is the vital truth beneath its catchwords: that though men cannot be equal, their access to educational opportunity can be made more nearly equal." [243]

The Massachusetts Institute of Technology (MIT). MIT's OpenCourseWare (OCW) Project had a bold aim: to offer the university's content to faculty, students and other learners at any time, for free. Since October 2003, the site has been reporting close to 16 million visits, split between new users and repeat visitors. OCW has grown to become an organization involving over 200 higher education organizations and associated institutions across the world. Member organizations span the globe and include Kabul Polytechnic University in Afghanistan, three universities in Columbia including Universidad Nacional de Columbia (Columbia), 19 universities in Japan including Kyoto University, the Institute for Social Sciences and Humanities in Russia, 32 universities in Spain, and 22 higher education institutions in the U.S. as of 2009.

The University of Texas offers many degrees mostly or entirely online through the "UT System TeleCampus." [244] Founded in 1998, the campus works with all 15 UT institutions to build and deliver online courses and degree programs. Students studying online can choose from over three dozen online degree and certificate options. While most students are from Texas, most U.S. states and more than 30 countries are represented in the student body. Students range in age from 17 to more than 65-years old. While the overall number of students online is small in comparison to other schools, its growth since its inception is impressive. Starting with 198 students in 1998, the TeleCampus had over 5,000 students in 2007.

General Electric Company. Many corporations create their own electronically based "corporate universities" to deliver education and training worldwide. G.E. invests over $1 billion in training and development for its employees worldwide and provides a global network of online learning with nearly 3.4 million online courses completed in 2007. [245]

Procter & Gamble: Procter & Gamble is a vast corporation with 138,000 employees in over 80 countries. In 2000, Procter & Gamble launched Rapid Learn, its virtual university, which introduced the concept of "surgical learning." According to this concept, employees learn only the particular areas of a technology or program that are specific to their job. When Rapid Learn was launched, 1,044 students downloaded web-based training (WBT). In 2007, the number of students downloading Rapid Learn WBT courses exceeded 50,000 a month. [246] A report from its manufacturing department estimated that its surgical learning methodology in Rapid Learn saved the company $75 in training for each employee download versus classroom and accelerated the pace of necessary high impact training by three to four years, according to an article in *Chief Learning Officer.* [247]

U.S. UNIVERSITIES TAP THE MARKET

More than two thirds of all higher education institutions now have some formal online offerings, with most of these providing fully online programs, according to Sloan-C. [248] As mentioned, as of Fall 2006, about 3.5 million students were taking at least one online course in the U.S, and nearly twenty percent of all U.S. higher education students were taking at least one online course. [249] Two year associate's institutions have the greatest growth rates and account for over one half of the online enrollments for the last five years, according to *Online Nation*, a report by Sloan-C. [250]

At the New York University School of Continuing and Professional Studies Online, students experience a state-of-the-art online classroom that lets them learn on a schedule that suits their lifestyle. Real-time online meetings with instructors and fellow students help turn your desktop into a classroom community. NYU-SCPS Online offers graduate programs, as well as continuing education courses and certificate programs. Courses are of equal quality, rigor, and credentials as New York University's highly acclaimed on-site programs.

And corporations are turning to universities for their online needs. For instance, Ingersoll Rand was looking for an MBA for its managers and chose the Kelley Direct MBA from Indiana University's Kelley School of Business. The course could be customized to fit Ingersoll Rand's needs, such as offering a course in global supply management, according to the *Financial Times*. [251]

THE GLOBAL COURSE PROTOTYPE

Creating the content for electronically delivered international courses is no easy task. Although worldwide use of the Internet is increasing at a dizzying pace (Table 1), cultural barriers in different areas of the world continue to make development of electronic coursework content a challenge.

Most courses are simply taught in the language and within the cultural framework of the country from which the course emanates. Others are modified for local audiences. There are courses, however, that were designed from the outset for an international audience. One particular course comes to mind.

In the 1990s, the Annenberg School of Communications and the Corporation for Public Broadcasting (CPB) joined forces to fund and distribute an electronically delivered course known in the U.S. as "Inside the Global Economy. [252] A group of educators and public television broadcasters from several countries met and collaborated on the video-based course. Besides the Annenberg/CPB group, the TELEAC Foundation of The Netherlands, the Swedish Educational Broadcasting organization, JL Productions of Chile, and the Australian Broadcasting Corporation were partners in the effort.

A 13-lesson video course was the outcome. Each lesson is based on an examination of two case studies. In addition to the 13 one-hour videos, the course requires a textbook in international economics, a course reader — created as a study guide for the course — and a software-based tutor that includes a glossary of terms, graphical analysis of data, interactive testing, forecasting simulations, and a databank of test questions keyed to the textbook chapters.

The course took three years to develop, was filmed in 20 countries, and was edited to fit the needs of each region in which it was shown. As this undertaking illustrates, creating content for a worldwide audience is not a simple task. But experiments such as "Inside the Global Economy" can teach us much.

The Internet takes this kind of curriculum one step further. Because of the World Wide Web's immediacy, using the Internet to deliver courses such as "Inside the Global Economy" enables content to be constantly updated, reflecting ever-changing world economic conditions.

THE ROAD AHEAD

The road ahead for electronically based international corporate education and training is still a bit bumpy, but it is definitely paved with a mix of silicon, fiber, satellite and mobile technology. It is a road that corporations will travel. As I've noted, they have no choice. In a shrinking world, where global competition is truly inescapable and where market economies rule, businesses worldwide must keep each one of their workers current with cutting-edge management and technical training. The communications revolution continues to fuse together with education in a very real way.

Businesses can do it themselves; they can contract with universities, contract with consortiums, or get involved in customized educational development partnerships. But they must train. They must educate. They must do it whether their employees are in Chicago or Calcutta. The key to success is finding the most cost-effective, flexible way to do it.

CHAPTER EIGHT
Principles of Good Practice

"Some people are willing to push the envelope of education delivery. And, when they do, accreditation agencies will be there to validate all or part of it."

Jack Allen, Commission on Colleges,
Southern Association of Colleges and Schools

In the U.S., several higher education groups have developed what are called principles of good practice in electronically delivered distance education.

Underlying many of the best practices are principles developed in an important article. In 1987, noted higher education Professors Arthur W. Chickering and Zelda F. Gamson outlined the seven principles for good practice in undergraduate education. Several hundred thousand copies of that article have since been distributed. It began by describing a harsh view of the higher education system in an article in *The*

American Association for Higher Education Bulletin: "Apathetic students, illiterate graduates, incompetent teaching, impersonal campuses—so rolls the drumfire of criticism of higher education."[253]

Good practice in undergraduate education, the article continued, follows seven basic principles. It encourages contact between students and faculty, develops reciprocity and cooperation among students, encourages active learning, gives prompt feedback, emphasizes time on task, communicates high expectations and respects diverse talents and ways of learning.

Since then, the Internet has emerged; and educators have sought to apply these principles to an online environment. In fact, the TLT Group, a non-profit assisting college and university educators, has set up a whole web site on the topic of applying the seven principles to the online environment. (http://www.tltgroup.org/seven/home.htm)

A whole host of other groups have developed principles and standards for online learning. Here is a sampling

- The U.S. Distance Learning Association developed ten principles of best practice for distance learning and 55 evidences of compliance found at http://www.usdla.org/accreditation3.php#4b. The ten principles relate to areas such as mission, standards, integrity, student enrollment and admissions, human resources, and the learning environment.

- Sloan-C has established five pillars of quality online education. They are:
  1. Learning effectiveness
  2. Scale
  3. Access
  4. Faculty satisfaction
  5. Student satisfaction. There are many sub categories within that.

On its web site, Sloan-C allows users to search by "context" based upon their perspectives (roles) in online learning, organizational type or subject area domain. Or they can search by the type of technology category (audio, video, synchronous, asynchronous, mobile, virtual and digital resources.)

In the U.S., two other higher education organizations have developed what are called principles of good practice in electronically delivered distance education. They are the Western Interstate Commission for Higher Education (WICHE) in Boulder, Colorado, and the American Council on Education in Washington, D.C.

The 17 WICHE principles (found at www.wiche.edu) were three years in the making and are called the Principles of Good Practice for Electronically Offered Academic Degree and Certificate Programs. They are essential reading for students considering enrolling in distance education programs and for institutions interested in offering such programs. They stress the development of rigorous educational outcomes, completeness of programs, adequate and appropriate interaction between students and teaching faculty, appropriate support systems and training for students and faculty, course and student evaluations, and the importance of students' access to learning resources.

The American Council on Education's principles for distance learning went into greater detail, but generally consisted of five over-arching concepts:

- That distance learning activities are designed to fit the specific context for learning;
- That distance learning opportunities are effectively supported for learners through fully accessible modes of delivery and resources;
- That distance learning initiatives must be backed by an organizational commitment to quality and effectiveness in all aspects of the teaching and learning environments;
- That distance education programs organize learning around demonstrable learning outcomes, assist the learner in achieving these outcomes, and assess learner progress by reference to these outcomes (see appendix for further discussion on assessment); and

- That the provider has a plan and infrastructure for using technology that supports its teaching or educational goals and activities.[254]

The council suggests, and I concur, that its principles of good practice should be applied beyond higher education institutions. All those involved in the learning enterprise — including individual learners, institutions, corporations, labor unions, associations, and government agencies — will benefit from principles that provide guidance in producing high-quality education with outcomes that can be clearly assessed (see appendix for a detailed example). Strengthening one sector will improve the effectiveness of the others and, in turn, address the learning needs of individuals and society as a whole, the council's task force said.[255]

ACCREDITATION: WHO CONFERS IT?

Accreditation historically has been the way for students to determine whether the institution they attend maintains certain quality standards. In most countries, the central government accredits higher education institutions. In France, for example, the national government is the accrediting agency. In the U.K., the national government essentially supervises accreditation. In Canada, the provinces are responsible.

In the U.S., the accreditation system works differently. Its different structure may make it a potential vehicle for certifying quality in electronically delivered international distance education. A look at the history of U.S. higher education accreditation provides some evidence for this notion.

While education in the U.S. has never been the responsibility of the federal government, in recent years, there are increasing resources from the federal government for elementary, secondary and higher education. The American Recovery and Reinvestment Act of 2009 (the so-called economic stimulus legislation) provided more than $100 billion in education funding and college grants and tuition tax credits, as well as billions more for school modernization. As Secretary of Education Arne Duncan said, "the primary goal of the stimulus is to save jobs — but the larger goal is to drive a set of reforms that we believe will transform public education in America. The four issues are: higher standards, data systems, turning around underperforming schools, and teacher quality." [256]

In the 18th and early 19th centuries, the states did almost nothing about educating their populations. Even though the first public school in the North American British colonies was created by local edict in Massachusetts in 1639, education primarily was the province of private institutions. It was delivered through an intricate web of private preparatory schools that prepared students for private universities. Because of its exclusive nature, education was reserved for the upper classes.

The situation changed as the country expanded to the West and its citizens became more mobile. At the end of the 18th century, in the Northwest Ordinance passed by Congress in 1787, the federal government encouraged the founding of universities on its frontier — in wild and untamed places like Michigan. And in the first half of the 19th century, public primary- and secondary-level schools proliferated as a result of decisions by states to fund them through state taxes.

Because the frontier universities had no private preparatory school system from which to garner students, they were faced with what can only be described as a recruiting problem. The dilemma was this: How could they convince students to continue their education at a university?

The solution was not long in coming from the University of Michigan through a program called the high school visitor plan. University faculty would visit public high schools in their state and examine both students and faculty to determine whether the school's students were "university material." The Michigan Plan caught on, and soon university faculties throughout the Midwest were visiting high schools.

The practice became so prevalent that by the early 1900s high schools were complaining they were being overrun by state university faculty. In mock desperation, the high schools suggested the process be turned around and that colleges and universities be the institutions examined. In fact, that is what happened.

During the first half of the 20th century a series of non-governmental, mostly volunteer, peer- review regional higher education accreditation agencies came into prominence. There are six for higher education: the Middle States Association of Colleges and Schools, the New England Association of Schools and Colleges, the North Central Association of Colleges and Schools, the Northwest Association of Schools and Colleges, the Southern Association of Colleges and Schools, and the Western Association of Schools and Colleges.

WORLDWIDE QUALITY STANDARDS

Because U.S. higher education accrediting bodies are nongovern-mental, some of their officials believe the U.S. could become the center for accrediting higher education programs worldwide. "We may lead the way because we are not government-affiliated. That offers an edge to the U.S. We can maneuver across international boundaries without competition. We also are able to transcend politics—if that's possible," contended Dr. Jack Allen of the Southern Association of Colleges and Schools.

The Southern Association already accredits universities in Latin America. Among those with its stamp of approval are the Monterrey Institute of Technology—with multiple campuses in Mexico and satellite courses that can be beamed almost anywhere in North America—and the University of the Americas, with campuses in Puebla and Mexico City, Mexico. The University of Monterrey in Monterrey, Mexico, also is an applicant for Southern Association accreditation.

WHAT ACCREDITORS THINK

Most accreditors in the U.S. agree that the fundamental principles of quality they use to judge traditional education institutions apply to electronic distance education institutions as well. For example, the integrity of an institution's conduct in all its activities, honesty and accuracy, adequate financial resources to run programs, and the like are applicable to both traditional and distance education institutions.

But accreditors admit certain characteristics of distance education make it unique and present challenges for distance education institutions and accreditors alike. One challenge is to develop methods for determining whether students and faculty are sufficiently computer-literate to either instruct or successfully complete a Web delivered course. How does the institution know, for example, if a student enrolling in an online degree program in business management is sufficiently knowledgeable in the technology used to deliver the courses to take exams? One could argue that the same question might be asked at a traditional higher education institution about the ability of a traditional student to understand how to use the on-campus library. It's normally not an admissions requirement, but most students get the hang of it rapidly.

Other issues include student access to library resources and arrangements for students to complete curriculums that are dropped from electronic institutions. Many accreditation agencies require traditional universities and colleges to provide an independent study option or the option of completing the curriculum at another university. It would not be difficult for a cyberschool to do the same.

And what about institutional visits by accreditors? "How do you visit an individual computer?" the Southern Association's Jack Allen quipped.

These are not insurmountable issues. The key is for distance learning institutions, be they in the public or the private sector, to be aware the requirements and provide for their fulfillment.

The North Central Association of Colleges and Schools (NCACS) has perhaps had the most experience dealing with the unique qualities of distance learning programs. Since the 1970s, it has accredited various forms of electronically delivered degree programs from universities including National Technological University, the University of Phoenix, The Union Institute and Capella University. And, as has been detailed elsewhere in this book, in 1999 the NCACS in a ground-breaking decision granted accreditation to totally online Jones International University.

Accreditation agency officials in the U.S. believe that the trend toward technological delivery of higher education will continue. "Its validation by these agencies affirms the contribution of distance learning to our economic and social well-being. It is providing increased access to quality education for students of all ages wherever they live and work," said Dr. Milton Goldberg, the former Executive Vice President, National Alliance of Business and the former Director, Office of Research, U.S. Department of Education. I strongly concur.

CHAPTER NINE
Cyberschools and You

"Truth, crushed to earth, shall rise again;
Th' eternal years of God are hers; But Error, wounded,
writhes in pain, And dies among his worshippers."
William Cullen Bryant[257]

A HOW-TO GUIDE FOR DISTANCE LEARNERS

If you've decided that distance education is an option you would like to explore in furthering your education or training, there are several important facts to keep in mind as you look for courses of study that meet your needs.

DISTANCE EDUCATION, NOT INSTANT EDUCATION

First, distance education is not instant education. For the majority of for-credit courses delivered via technology, it is necessary to submit an application for admission to the program and arrange for academic records to be transferred from other educational institutions you have attended. After that, you must, of course, be admitted to the distance education course or program you have applied for. Even though many cyberschools will expedite this process, to gain admission to a distance education institution, you follow much the same process as you would to gain admission to a traditional institution. If you are applying to take courses via distance education at a traditional university—for example, Pennsylvania State University in University Park, Pennsylvania, has a large distance education department—you must meet its admission standards.

A RIGOROUS WAY TO LEARN

Second, distance education is rigorous. Don't expect coursework to be easier simply because it is delivered via technology. Indeed, students must not only have the ability to absorb and understand the courses' content but also be disciplined self-starters. To an even greater extent than on a traditional college campus, no one will stand over your shoulder admonishing you to get your work done. It's up to you. And it should be. After all, you're the one getting the education.

CYBER EDUCATION, NOT PASSIVE EDUCATION

Third, when you are a cyberstudent, it's virtually impossible for you to be a passive learner. The approach to learning is active, not passive. In traditional classrooms, students can get by for weeks at a time sitting quietly taking notes—or not taking notes—and never actively participate in discussions with the instructor and other students. Many of us have done this at some point.

In a cyberschool, students nearly always must react or provide some appropriate input to continue to the next phase of an assignment. In addition, courses are more often designed to be collaborative efforts among students than are traditional lecture hall courses.

In some cyber courses, a significant part of a student's grade depends on participation in online discussion groups.

A BETTER WAY TO LEARN?

Fourth, learning via technology may be a better way for some students to learn.

In its Apple Classrooms of Tomorrow project, Apple Computer, Inc., investigated how teaching and learning change when people have constant access to state-of-the-art technology. The project began in 1985 and is ongoing. After the first 10-years, the project's conclusions were that students become re-energized and more excited about learning when using information technology. Grades improved, standardized test scores went up, and dropout and absenteeism rates decreased, according to Apple's report on "Teaching, Learning and Technology." The study reported that during the first 10-years of the project, dropout rates for participating high school students fell from 30 percent to zero. [258]

Even before the Apple project's 10-year findings, it was reported time and again in research on how we learn at all educational levels that electronic instruction, either via teleconference or computer conference, can be as effective as traditional classroom-based lectures and face-to-face discussions. Online students have test scores equal to those of students in conventional classrooms if the quality of the teaching is the same, they report better access to instructors, and they reportedly improved their ability to collaborative efforts among students that are traditional lecture hall course.

The follow-up to Apple Classrooms of Tomorrow is the Apple Classrooms of Tomorrow — Today. It sets out six design principles for the classroom of the 21st Century: an understanding of 21st century skills and outcomes, a relevant and applied curriculum, an informative assessment, a culture of innovation and creativity, social and emotional connections with students, and ubiquitous access to technology. [259]

EIGHT QUESTIONS TO ASK

Knowledge is power. It's a good idea to ask any distance education institution you are considering lots of questions. Here's a list of eight to get you started.

1. **How technologically savvy do I need to be to take electronically delivered distance education courses?** It almost goes without saying that for distance learning courses delivered via technology, some degree of computer literacy is necessary. Usually a cyberschool will give you a list of the technologies that must be accessible to you when you apply for admission. Just in case a list isn't offered, ask about hardware, software, computer, and peripherals requirements. Also, broadband access is usually preferred.

2. **What kind of student support structure does the institution offer?** Is there a way to contact someone about course advising, missed assignments, student emergencies, or any other question that might come up while you're taking courses at a distance? Is there a number to call? Is it an 800 number? What is the email address? What social media (wikis, blogs, etc.) are available? What kind of library support services are available? Find out.

3. **Is the institution accredited?** This question is important for several reasons. First, and perhaps foremost if you are hoping to receive federal financial aid to attend classes at a cyber-school or at some less high-tech distance education institution, it's important to know that the U.S. government does not lend money to students attending non-accredited institutions. Similar restrictions apply in other countries. In addition, though accreditation is not absolutely necessary for a school to stay in business and award certificates and degrees, a degree or certificate from an accredited institution is far and away more prestigious for a student. Primarily because they are so new, many cyberschools are going through the accreditation process now, and their applications are being carefully scrutinized. According to accreditors

in the U.S., technologically delivered education is legitimate, and if it's not already recognized by a particular accrediting agency, it soon will be. For more information about accreditation, see the previous chapter.

4. **Where do I buy the textbooks required for the course?** Textbooks still are part and parcel of any coursework. Make sure you know how and where to obtain yours.

5. **Are the instructors at the cyberschool trained to teach online?** This is a very important question to ask because teaching in cyberspace requires a different, more interactive approach than teaching students in a classroom.

6. **Is it possible to complete an entire degree program via technology through this institution?** To effectively plan your educational strategy, you need to know how much coursework you can complete online or via other distance education methodologies.

7. **How long does it take to get a degree or certificate through this institution?** The answer may be the same amount of time it takes to complete a course of study at a traditional university, or it may be less. Often the timing is flexible, because one aspect of good distance education is that it allows the student to be in charge of the process. There also may be minimum and maximum allowable times. Check out your options.

8. **What if I start a degree program and find out when I'm in the middle of it that the program has been discontinued?** The majority of degree-granting institutions — whether they are cyberschools or traditional universities — make provisions for alternative ways to complete degree programs that have been dropped. Such provisions are required for an institution to be accredited. Find out what alternatives the institution in which you plan to enroll can offer you if it discontinues your degree program.

CYBERSTUDENT OPINIONS

In addition to asking questions, another excellent way to judge the quality of a cyberschool or other type of technology-delivered distance education institution is to listen to those who have gone through a program before you. As I've mentioned throughout this book, students enrolled in cyberschool courses represent diverse backgrounds, a broad range of ages, and widely varied goals.

Many are striving to enhance their career opportunities or to attain other personal and professional objectives. Some audit courses simply out of curiosity or a commitment to lifelong learning. All cite distance teaching's focus on the student as a primary factor in their decision to take advantage of this nontraditional way to learn.

The following are several profiles of distance education students that might prove enlightening. Though the profiles by no means represent the varied and plentiful distance learning programs available around the world, I hope they will help illustrate my point that the marriage between learning and communications technology is living up to its promise as a way to make all the world a school, one that millions more can afford.

Corey Christiaens completed a Master of Arts in Education/Administration and Supervision at the University of Phoenix. "When I decided to pursue my master's in administration online with University of Phoenix," he said, "I was concerned that the lack of face-to-face time in the classroom might make it difficult to have an authentic learning experience. But my learning team experiences changed that. One of the greatest things about the program was working with a global mix of people from all over the world. It added a great perspective and dynamic to the class." [260]

Judith Church, who received a Doctor of Health Administration at the University of Phoenix, said her "University of Phoenix doctoral studies helped me learn to combine abstract and concrete ideas, as well as search the literature for support of my articulations. I was exposed to business and finance theories, enabling me to view the health care field from a broader perspective. I also learned from the expertise of my faculty and learning team members. It was an awakening experience."[261]

Samuel Brackeen, who received his Master's Degree with dual concentrations in Business Management and Project Management from Colorado Tech Online, said, "My learning experience with Colorado Tech Online was just phenomenal. The experience has opened up doors for me in many professional areas. Without Colorado Tech Online I would not be where I am today in my career. The main thing... that captured my attention was the ease of communication between the students and the instructors. Students can communicate with their instructors easily and frequently."[262]

Oya Bozkurt, in her third course at Jones International University, has been an elementary school teacher for over 10-years. She has taught all grade levels from K–6 and currently is teaching third grade. "When I think back to the start of [my first course], I was a different person. I was shy, withdrawn and lacked effective educational knowledge and communication skills I didn't want to be around people, because I didn't know whether or not my methods of teaching were right or wrong. Simply put, I did not have names for what I was doing and the reasons for doing it. Now, I can teach with confidence. I have learned to use the proper terminology when communicating with colleagues and administrators. I now know what signs/cues to look for in my students when I am teaching so that they are learning effectively. I am now aware of my actions, and more so my choice of words to use with parents, and school officials but most importantly, students. I now know that I can contribute to new solutions and offer a variety of methods to relieve my own frustrations and also my colleagues' frustrations about teaching and the various disciplines of it. I have learned to tackle life's hardships one at a time. And, most importantly I have learned that I must accomplish this program if I am going to be an educator. Not just a teacher but an educator, someone who sees the abilities, capabilities, and skills that each child brings with them into the classroom. I cannot believe that I have changed so much over these past eight weeks."

Derrick Percival had been in Department of Defense procurement for 12 years and said that through Jones International University he has "seen what the selling side of industry has to deal with. Now, I am learning what it takes to establish a business, build it up, and be successful. This course is challenging and makes me think 'outside the box', which I am enjoying. If there is one thing that I have learned from my past, never take anything for granted, you never know where you are going to end up. Not only will obtaining my degree allow for possible advancement, it could allow for other "doors" to be opened that would have never been possible had I not gone back to school."

Deborah Sakarov is in the distance master's degree program at Indiana University and was working to get a master's degree in Instructional Systems Technology from Indiana University at press time. She completed all the required courses and had 2 electives to go. An instructional designer and developer of a variety of online education-training projects for business and government, she said, "In the really good fun and challenging online classes, a lot is going on all the time: from virtual discussions between students, to YouTubes, video lectures, and readings. The best classes are the ones where the professor is accessible; where if you have a question you can email or chat, and can get an answer when you need it. An online class can be more lively and include more homework than an in-person course. We have books to read, weekly assignments, collaborative projects, and several online discussions where we participate nearly every day. Not only do we discuss, but we argue."

"Some online classes, especially the ones that Deb Sakarov enrolled in, are magical experiences," said Curtis Bonk, Professor of Instructional

Systems Technology at Indiana University. "Adult learners come to learn and you, as the designated instructor, find it difficult to close down a single discussion post, let alone the class at the end of the semester. Adult learners are often energized by each other. They share, participate and enjoy each other. The endless trails of insights and experiences from each other, to each other, are inspiring to say the least. YouTube, Scribd, Facebook, Wikibooks, Google Docs, Blogger, and so on, have arisen the past few years to turbocharge an already exciting process. Sure it takes some planning skills on part of the instructor as well as heavy doses of leadership, persistence, and structure, but the rewards are immense. There is nothing like it. You can teach any student located anywhere on the planet at any time of day. The world of education is opening for online learners and instructors before our eyes."

Rod Hicks envisions a time when patient care errors are a thing of the past. He's putting his passion for public health, his research in patient safety, and his Capella University PhD to work to improve the health care system. As the nation's first endowed chair for patient safety in a school of nursing—a position created through a partnership between University Medical Center (UMC) Health System and the Texas Tech University Health Science Center in Lubbock—and as a full professor of nursing, Hicks will promote a culture of patient safety for nursing students and the clinical staff at UMC.

Hicks became a nationally recognized leader in patient safety while employed at the U.S. Pharmacopeia, where he managed the Patient Safety Research Division. "I evaluated medication studies and gained a very strong knowledge of patient safety issues at the national level. That experience and my background in nursing left me very qualified for this endowed chair position."

When Hicks decided to earn a PhD, he was drawn to Capella because of the flexible online format and the university's focus on adult learners. "Capella shaped my thinking and recognized my goals and values as an adult learner, encouraging a synergy between my professional responsibilities and academic opportunities. Without the synergy, I wouldn't have been able to publish 75 articles by the time I graduated."

Hicks already has several major initiatives underway. He is working to formalize a transdisciplinary Center for Excellence in Patient Safety, comprised of health professionals from medicine, nursing, pharmacology, and allied health fields at Texas Tech University Health Science Center and two local medical centers. He is also collaborating on a program to develop Patient Safety Coaches who can monitor patient care protocols and identify potential for errors. Today, Hicks has published more than 120 articles and is enthusiastic about giving back to his profession and his community. "I'm following Capella's collaborative mentor-scholar model to support others on their career path and contribute to the body of knowledge for my profession." [263]

CHAPTER TEN
Free Market Fusion: One Path

*"The whole of science is nothing more than
a refinement of everyday thinking."*

Albert Einstein in Physics and Reality[264]

The electronic delivery of education is an ideal prospect for a kind of public/private partnership I call Free Market Fusion, a management process I have studied for several years. It was chosen as the final topic for this book because it offers students, teachers, and public and private education planners an additional way of thinking about how they might find the resources to pursue electronic education.

My book, *Free Market Fusion: How Entrepreneurs and Nonprofits Create 21ˢᵗ Century Success* (Cyber Publishing Group, Inc., 1999), treats this subject in more depth and explores several case studies. A companion document, with interactive exercises, is available on the Web at www. freemarketfusion.com.

DEFINING FREE MARKET FUSION

What is Free Market Fusion? It is an entrepreneurial approach to identifying or creating opportunities for innovative solutions. In physics, fusion occurs when two elements combine to create a new element and, simultaneously, release a tremendous amount of energy. Rather than converting one form of energy into another, the reaction instead creates new energy. Fusion is one of the most powerful and energy-efficient processes known to the world; we race to find a means to safely harness its potential in the service of humanity.

Free Market Fusion is both a process and its result. It is a process that creates new products, services, and solutions and is typically based on new or modified concepts. Although portions of entire industries can be involved and permutations of participants can range widely, it is easiest to discuss the process by considering first a few examples that most clearly demonstrate the process and potential of Free Market Fusion.

Free Market Fusion is the coming together of two or more entities, one or more of which is characterized as a for-profit enterprise and one or more of which is characterized as an institutional, nonprofit, quasi-governmental, or governmental entity. For purposes of illustration, we'll call them A entities (for-profit) and B entities (institutional, nonprofit, etc.). The process typically culminates in the fusing of a portion, or possibly all, of the assets of one or more A entities with a portion, or possibly all, of the assets of one or more B entities.

Although Free Market Fusion may result in the formation of new enterprises, typically at the outset, existing organizations are the creators. Also, typically the entities involved share a common need, concern, or opportunity that generates support for resolution. The collaborative process inherent in Free Market Fusion can engender the tremendous release of energies that comes from looking at the world not as a miasma of intractable problems but as an arena of challenges awaiting exploration, initiative, and solution.

WORKING TOGETHER

In the Free Market Fusion process, each entity contributes its particular strengths, agreed upon by both parties, to the project. For example, in a partnership between an entrepreneurial group and an institution, the entrepreneur might contribute the initial innovative idea as well as technological marketing expertise and significant risk assumption. The institution might contribute personnel, physical facilities, familiarity with the existing market, and perhaps acceptability.

Depending on the parties, some of those roles might be reversed. However, the purpose of the partnership is always to enable both parties to accomplish goals neither could attain alone to create a solution where there was none. As in fusion, the new solution is accompanied by a breathtaking burst of energy as new possibilities and opportunities open up to everyone involved, both those creating the solution and those benefiting from it.

The university system in the U.S. overall is a great example of the fusion between the public and private sectors. As The Economist pointed out, "America has pioneered the art of forging links between academia and industry. American universities earn more than $1 billion a year in royalties and license fees. More than 170 universities have business incubators of some sort, and dozens operate their own venture funds." [265]

FREE MARKET FUSION, ENTREPRENEURS AND INSTITUTIONS

A productive type of Free Market Fusion results from combining the strengths, resources, and assets of an institutional entity with those of an entrepreneurial group. In this situation, the strengths of the institution might include specialized subject knowledge, existing facilities, thorough understanding of a specific market, a strong administrative and management structure, and a history of solid, stable functioning.

The strengths and assets of the entrepreneur might include expertise in competitive strategies, the ability to evaluate risks and a willingness to undertake intelligent ones, commitment to innovative thinking, awareness of opportunities presented by recent technological advances, strategic networking and distribution abilities, an understanding of and familiarity with capitalization resources, and the ability to orchestrate the transformation of concepts into products.

The catalyst is leadership, which may or may not be provided by the freewheeling entrepreneur or company. The fact that an entrepreneur may be well known and accustomed to operating in the public spotlight does not mean that person will assume the leadership mantle or that he or she should.

Leadership encompasses much more than simply assuming the role of primary public spokesperson. The most critical leadership activities are intuition, imagination, planning, organizing, networking, and acting as missionary within the organizations involved, persuading and recruiting internal supporters for a new concept. Often individuals with established credibility within an institution can do this most effectively. The entrepreneur may assume some parts of this role or merely advise and be an "outside" networker, promoting the concept to other organizations and individuals whose support is essential. Even the process of leadership may be shared or fused. Entrepreneurs and institutions provide an especially effective example of Free Market Fusion because combining many of their core strengths enables them to accomplish what neither could accomplish alone.

INSTITUTIONS: INERTIA VERSUS INITIATIVE

Institutions are a critical part of society's infrastructure. They include schools, colleges and universities, hospitals, prisons, the military services, national charitable organizations, unions and professional organizations, quasi-governmental agencies, and such community entities as libraries, symphonies, museums, civic leagues, and innumerable religious groups.

Often, they have existing physical facilities and a stable organizational structure. Successful institutions have a thorough understanding of their constituencies and of those constituencies' special needs and concerns. Often they bring the comfort of market acceptance occasioned by their involvement in the new solution.

Institutions play an important role in reaffirming our sense of community, especially today when we are deluged with an onslaught of change on a regular basis. As connections to our past, they are familiar and comforting. Many of them have existed almost as long as the country itself; others grew with the needs of our growing nation. Harvard was established in 1636, Yale in 1701, and in 1862 the Morrill Act led to the establishment of the public higher education system. As of 2007-2008, the U.S. had over 4,300 institutions of higher learning. [266]

The first lending libraries in the U.S. were founded by English clergyman Thomas Bray in Maryland in the late 1600s, and the country's public library system was launched nationally when Andrew Carnegie undertook funding the construction of 2,500 community library buildings between 1881 and 1891. Over 9,000 public libraries of all sorts are now supported by communities across the U.S.

The Boy Scout and Girl Scout programs, originating in Great Britain, were introduced in the U.S. in the first decade of the 20th century and now involve millions of children, teens, and adults. The YMCA, with 45 million members in more than 124 countries, has been a pillar of thousands of communities since its inception in London in 1844. And U.S. towns and cities have relied on community hospitals ever since Philadelphia's Pennsylvania Hospital first received its charter in 1751 through the tireless efforts of Benjamin Franklin.

Institutions have played a central role in advancing the goals of society throughout history. It is imperative that they remain as vital and forward-thinking as possible if they are to continue their positive impact on society. This is no easy task; it is in the nature of institutions (and of monopolistic businesses) that stability may deteriorate to stagnation and management to mediocrity. Even Thomas Jefferson recognized this possibility, when he wrote:

> "I am not an advocate for frequent change in laws or constitutions. But laws and institutions must go hand in hand with the progress of the human mind. As that becomes more developed, more enlightened, as new discoveries are made, new truths discovered and manners and opinions change, with the change of circumstances, institutions must advance also to keep pace with the times. We might as well require a man to wear still the coat which fitted him when a boy as civilized society to remain ever under the regimen of their barbarous ancestors." [267]

Over the past several decades, we as a society seem to have lost confidence in the ability of our institutions to perform with competence and integrity the functions for which they were created. As circumstances have changed, often institutions have failed to change with them, choosing instead to hold onto the more familiar, less-threatening solutions of yesterday and to be protected by the environment that depended on them. The tendency of institutions and large organizations to rely on solutions drawn from yesterday's realities was pointed out some two decades ago by Peter Drucker in his article "Managing the Public Service Institution:"

> "No success lasts 'forever.' Yet it is even more difficult to abandon yesterday's success than it is to reappraise failure. Success breeds its own hubris. It creates emotional attachments, habits of thought and action, and, above all, false self-confidence. A success that has outlived its usefulness may, in the end, be more damaging than failure. Especially in a service institution, yesterday's success becomes 'policy,' 'virtue,' 'conviction,' if not indeed 'Holy Writ,' unless the institution imposes on itself the

discipline of thinking through its mission, its objectives, and its priorities, and of building in feedback control from results over policies, priorities, and action."[268]

RISK-TAKING: THE KEY ROLE

A complementary relationship between entrepreneurs and partnering institutions relates to risk-taking. Missteps within a institutional environment can easily spell the end of a promising career, a circumstance that has an obvious (and understandable) dampening effect on an institutional leader's willingness to take risks. In addition to identifying opportunities, then, another of the entrepreneur's key roles in a Free Market Fusion venture is to assume a substantial amount of the risk involved in any new undertaking, thus diverting a large measure of the "exposure" from the institution's leader to the entrepreneur.

COMBINING RISK-TAKING AND CAUTION

This imposes no hardship, for although risk-taking is anathema to an institution, judicious and well-informed risk-taking is second nature to the entrepreneur. An entrepreneur has the freedom to respond to opportunity with a desire for gain rather than resisting it because of a fear of loss. Similarly, because entrepreneurs are not part of the "old guard" operating environment of the institution and have minimal vested interests in conforming to established ideologies, they are much freer to envision radical alternatives and innovative solutions outside the boundaries of accepted practices.

Thomas Jefferson believed that every generation needed its own revolution. In the U.S., entrepreneurs have always been society's revolutionaries, playing from outside the boundaries, creating new solutions for a changing world. Now they have an opportunity to chart new ground once again.

Exploring this new frontier will take discipline and commitment to a common vision, because the gains to be won through society's revitalization will be more long-term than immediate. No work is more critical, however, for the well-being of the world; we simply cannot continue to advance, or compete fairly in the global marketplace, if major portions of our population or our infrastructure are left to perish in the wasteland of yesterday's solutions.

MODELING FREE MARKET FUSION

The major components of any Free Market Fusion process are:

1. Identifying and evaluating potential Free Market Fusion opportunities;
2. Creating an innovative solution;
3. Identifying potential partners;
4. Structuring the relationship; and
5. Undertaking the project.

Obviously, every situation will demand different levels of time and energy at each phase. However, if participants know from the outset that there is a process to work through, then resources can be allocated accordingly.

If an innovative solution incorporates a fairly nontraditional concept, it will be easier to work with a partner who already is comfortable with the nontraditional concept. For example, in the late 1980s JonesKnowledge.com combined a nontraditional delivery process (cable television) with a nontraditional teaching method (telecourses). The colleges and universities that had not previously used telecourses were not nearly as likely to be comfortable with the concept represented by JonesKnowledge.com as were the schools that understood the potential of telecourses and had already integrated them effectively into their programs. The same applied to the use of Internet technology in the mid-1990s.

Another consideration is that many potential partners may be constrained by people or organizations whose vested interests might be threatened by the entity's move into a new arena or into a relationship with another (autonomous) entity in which the vested interests have no control. A major contributor to the organization, for example, may forbid it from entering into any new relationship for fear that the contributor will lose his or her tacit control of the organization's goals and direction.

This is a fairly predictable response. Fear of change is a familiar reaction, especially for constituencies, such as labor union memberships or government or large business entities, that fear they may lose previously protected positions. Therefore it becomes critical to strive for an acceptable level of friction, in which the fear of change is counterbalanced by enthusiastic commitment to the opportunity at hand. This control relationship may not surface initially, but when it does, it often terminates further negotiation.

STRUCTURING THE RELATIONSHIP

Once the concept of a Free Market Fusion venture is developed, then the manner in which these entities and their functions, equipment, personnel, or activities can be joined must be considered. What are the costs? Who must contribute what? Who might feel threatened? Who will manage the process? What kind of time frame will it take for Free Market Fusion to function? What are the risks involved and who will take them? What is the reward system? What defines success? The list of potential questions to be answered is long and will vary with each project.

Obviously, each project will have its own set of circumstances and concerns that need to be addressed and agreed on before other steps can be taken. However, the following areas can serve as a starting point from which to explore and negotiate.

Goal issues. What are the purposes and goals of this project? How will achievement of the goals be measured? How and when will they be evaluated? What is the reward structure? An agreement must be reached about what constitutes success for the venture.

Inertia issues. Best-laid plans can easily be derailed by organizational inertia. How rapidly will both parties be able to respond to opportunities or crises? How rapidly are both parties willing to respond?

Structural and logistics issues. How will the project be undertaken? Where and how will it be located — centralized with one participant, headquartered at a project site? Who will implement what aspects of the project?

What is a reasonable and mutually agreeable time frame? This can become a key issue if both parties do not understand and accept how long it will take to accomplish key tasks. If the project will entail working with large institutions, government agencies, or other bureaucracies, (including bureaucratic businesses), the time from start to completion could be substantially extended.

Competition issues. How will you deal with competing players? Will you work around their established programs, trying not to disrupt their "market share," or will you try to displace their "product" with your own.

Much care must be taken in dealing with societal concerns. Society is rarely damaged when, in the rough- and-tumble competitive market of consumer goods, a candy bar or a laundry detergent or a sports car line bites the dust. However, if an institutional partner fails, the impact on society can be quite substantial. When societal issues are addressed, often a less-than-terrific solution may be worth keeping because it provides ancillary benefits.

Long-term issues. Assuming the project is successful in meeting its goals and is profitable, what should become of it in the long term? Should the relationship between the participants continue as is, or should it be reviewed on a specified basis? Should the project continue in its current form or be taken over by one of the participants? Should it be taken public as an established company? Should it move into other Free Market Fusion arenas?

OPPORTUNITIES FOR FREE MARKET FUSION

There are many areas in which a Free Market Fusion approach currently is or soon will be enabling us to make more creative, effective use of the technology. A few examples follow.

Health Care. As the 2008 U.S. Presidential campaign clearly underscored, it has become clear over the past several years that the world's health care crisis is going to demand radical measures and innovative solutions. The twin goals of cost containment and universal access to basic levels of medical care will obviously remain mutually exclusive unless efficient, cost-effective alternatives can be developed and implemented rapidly. A Free Market Fusion process combining the strengths and knowledge of the medical establishment with the vision and technological savvy of entrepreneurs offers means of achieving these goals.

A scarcity of physicians in impoverished urban areas and in the world's geographically isolated rural areas makes delivery of even the most basic, preventive-medical care difficult if not impossible. But recent advances in two fields — remote diagnosis and home medical testing — are proving that quality health care and reasonable costs can go hand in hand using information technology.

Remote diagnosis, incorporating advances in computers and telecommunications, enables communities to avail themselves of state-of-the-art medical technological expertise well beyond the means of their local medical practitioner. In addition, when geographically remote or inner-city communities can avail themselves of these technologies only as necessary, they can focus resources on basic medical care provided by less costly medical professionals such as paramedics and nurse practitioners.

Like remote diagnosis, home health screening, medical testing, and diagnostic technology can save time, money, and lives. In addition, innovations in medication-delivery tools are enabling patients to self-administer oxygen, shots, and even intravenous food through a pump carried in a small nylon backpack. These products and services have been developed by medical entrepreneurs who saw better ways of meeting individual health care needs.

Environment. More than 39-years have passed since the first Earth Day, in April 1970, called the world's attention to the deteriorating state of the global environment. Since that first tolling bell of warning, we have become increasingly familiar with the challenges that confront us. Global warming and its greenhouse effect have continued unabated as waste gases, primarily carbon dioxide released by the combustion of oil, coal, and gas, continue to spew into the Earth's atmosphere. These same energy sources will one day be expended. Meanwhile, the world's waters and aquatic species are still being poisoned by acid rain, largely the result of sulfur dioxide released into the air by coal-burning power plants. The Earth-encircling ozone layer is less and less able to protect us from the life-threatening effects of the sun's ultraviolet rays as chlorofluorocarbons continue to eat away at this protective blanket. In tropical regions of the Third World, growing populations desperate for economic survival burn their forests to clear enough land to graze cattle or cultivate marketable crops, taking more and more tropical rain forests out of the increasingly precarious global ecological balance. The industrialized world's reliance on nonrenewable resources guarantees the ongoing acceleration of these frightening trends.

So is the world set for a so-called Malthusian disaster? *The Wall Street Journal* examined this issue. It pointed out that the world's population and prosperity have exploded and the "result is that demand for resources has soared. If supplies don't keep pace, prices are likely to climb further, economic growth in rich and poor nations alike could suffer, and some fear violent conflicts could ensue."[269] But the article concluded as follows: "Indeed, the true lesson of Thomas Malthus, an English economist who died in 1834, isn't that the world is doomed, but that preservation of human life requires analysis and then tough action. Given the history of England, with its plagues and famines, Malthus had good cause to wonder if society was condemned to a perpetual oscillation between happiness and misery. That he was able to analyze that 'perpetual oscillation' set him and his time apart from England's past. And that capacity to understand and respond meant that the world was less Malthusian thereafter."[270]

Technology has enabled a multitude of innovative environmental solutions across a broad range of targets. Alternative, renewable energy sources such as solar photovoltaic cells, geothermal and solar-thermal generation, wind power, and hydrogen are no longer dismissed as fringe thinking. "Industrial ecology" became the manufacturing credo of the 1990s, as more and more companies understood that by redesigning manufacturing processes they could avoid using materials that end up as toxic waste and thereby avoid the costs associated with disposal or storage of toxic material.

Sustainable development is a goal we all must embrace. It is imperative to accept the fact that market-based environmentalism offers the most effective means of transferring technological advances into the areas of greatest need, ensuring that future generations inherit a land, not a landfill.

As Malthus stated, "the power of population is so superior to the power of the Earth to produce subsistence for man, that premature death must in some shape or other visit the human race."[271]

CHALLENGES ARE PLENTIFUL

The list of challenges we face is long and daunting. A quick hit list of social concerns might include the high dropout rate for high school minorities, illiteracy, child poverty, drug addiction, crime, AIDS, homelessness, overcrowded prisons and a sky-high recidivism rate, the need to increase the effectiveness and accessibility (financial as well as logistic) of our higher education resources, to re-incorporate seniors into productive societal roles, to mainstream the physically disabled back into society so that they can live independent lives and make the contributions they are capable of, and to provide adequate, affordable medical care.

The world is caught in a web of suffering that decades of goodwill, foreign aid, and well-intentioned but often fruitless efforts have failed to eradicate. The world hunger problem and its companion issues of disease, is ongoing. As Malthus stated "population must always be kept down to the level of the means of subsistence". [272]

Where do we begin? How do we begin? We can begin with Free Market Fusion.

THE LARGER ARENA: TAPPING ENTREPRENEURIAL TALENT

Why not go to the world's entrepreneurs and ask them for solutions? These are individuals trained to see the opportunities in change, the possibilities in dislocation. Not as constrained by governmental structures or established processes, entrepreneurs are free to find the most effective ways to meet goals. Who knows what reordering of existing resources, what rethinking of current responses, we might achieve? We need to tap the creative energy and risk-taking spirit of those willing to operate in Buckminster Fuller's "outlaw area" of untried solutions and no guarantees.

I believe solutions are always possible. The key is to structure circumstances that nurture creative, innovative thinking so that our most innovative thinkers can design new solutions. It is obvious that when a society faces a problem that has continually resisted traditional means of resolution, other solutions must be invented and tried. It becomes necessary to think and create outside the structure of established assumptions and policies with great speed.

The challenges that have been described in this book and with which higher education must deal are formidable. What has become obvious is that the changes will occur. People who want additional training or higher education will find a way to get it and, with rapidly opening free markets, there will be suppliers.

The challenge to our world's higher education establishment is specifically to accept this inevitable shift in its student market, to identify partners and technologies that will help it respond, and to set about creating ways to become more effective educational suppliers in the 21st century. Clearly, cyberschools have a meaningful part to play.

CHAPTER ELEVEN
Epilogue

"'Tut, tut, child,' said the Duchess. 'Everything's got a moral if only you can find it.'"

— *Lewis Carroll, Alice in Wonderland*[273]

The information revolution has been largely consumed by the knowledge revolution. As some unnamed sage once quipped, "Successful revolutions hasten their own demise."

The knowledge age will be an age of intelligent machines and convergent beings who gradually come to depend on networks, computer chips and data banks to enhance their individual characters and capabilities. It will be revolutionary in impact and highly controversial. Technology will drive it. Second in importance to the changes brought about by a revolution are its artifacts, which represent the catalysts of change.

It has been observed by others that the valuable artifacts of the American Revolution were not its military paraphernalia but the colonial printing presses and the information they produced: Letters, newspapers, pamphlets, extracts of sermons and speeches, and, that singular document, the Declaration of Independence. Many of the documents predated the actual hostilities, or, in the case of the Declaration, clarified for posterity the passions that were unleashed after almost two decades of dissent and protest in the American colonies.

Also observed was that never in the history of humankind had political discussion and debate been so effectively or spontaneously communicated as in the outpouring of political rhetoric flowing from the colonial presses. Thousands of cheap leaflets and pamphlets were printed, passed hand-to-hand, and reprinted so that the 13 colonies were literally drenched with the ideas of democracy.

When future anthropologists sift through the artifacts of the information revolution and the beginnings of the knowledge age, they will come across the skeletal outlines of electronic platforms and devices of the late 20th and early 21st centuries. These made our swift transition into the age of knowledge possible.

The massive proliferation of information across multiple communications systems constantly became smarter and smarter, connecting incessantly evolving software with a larger and larger array of digital devices. The onslaught continues.

All of this is changing, among other things, the makeup of knowledge acquisition — especially teaching models, as they have existed since the time of early Athens. Pedagogy is being turned on its ear.

Substantial parts of the world's educational systems are being transformed into cyberschools. For those societies that are wealthy, the transition is coming almost overnight. For those that are poor, mobile technology and devices are allowing the transition to happen as well, and the accelerating wave of change is as inevitable as the wave of illiteracy eradication that has swept the world in recent years. To be sure, the battle has not been won, but the new technologies of the communications revolution, including the Internet and the tremendous power of search engines such as Google, along with their progeny and derivatives, will help propel it forward.

These are technologies of freedom. They can democratize many things, including education. They not only guarantee delivery but can also assure standards of excellence and act as a potent weapon against censorship and information control. They provide a level playing field, assuring that students who enter a cyberschool immediately have at their disposal vast resources of electronically stored and linked information resources that can quickly put them on a par with their contemporaries. These resources, combined with a willingness to manipulate them appropriately, also place the cyberstudent at the center of the education equation. Students can have access to the primary sources, unfiltered by the opinions of others. The cyberstudent is in the unique position of being able to question and challenge assumptions and hypotheses as never before.

The underlying powerful aspect of cyberschools is that they make access to educational opportunities more equally available to more people. They do so regardless of where people are, and in countless cases regardless of who they are or what their condition in life might be. Distance is no longer a barrier and even the barrier of time is partially erased.

As mentioned in the earlier Chapter Six on M-Learning, mobile wireless is allowing the world's poor to bypass traditional infrastructures and move right to wireless infrastructures. This will transform virtually every setting that has electric power—living rooms, one-room schoolhouses, and yes, even one-room huts—into access points for the age of knowledge.

Such delivery systems can be free of much of the friction of traditional education. Certainly quality assurance will remain an important part of institutional oversight, but hopefully accrediting organizations will continue to change the way they view students and institutions.

Cyberschools provided an augmentation to the world's educational systems in the late 20th century, but in the 21st century they will quickly emerge as a major economical solution to satisfy the increased global demand for education. Importantly, they can provide "scale" of delivery and wider market access for educational solutions at a more affordable price.

Still, no one existing institution, company, or industry can do it all; and the government can no longer afford to do it all. There is room for everyone to participate. And, to survive and evolve, participate we must.

For educators who are reluctant to accept electronic educational delivery because of its unavoidable assault on cultural customs, I offer the assurance of philosophers from McLuhan to Plato. As McLuhan observed, television — and let's extend this to cyberschools — created a new environment through which we observe the "old" environment of the industrial age, helping us to understand and learn from it. Likewise, the industrial age transformed the Renaissance into an art form because of the new perspective through which it could be viewed. Plato, the scribe of Athens when writing was new, turned oral dialogue into art by documenting it.

It is my conviction that the communications revolution and its related technologies, fused with education, is propelling us into a new renaissance — the knowledge age. This age will accelerate as the private side of our world economy and cultural life helps to tackle the world's great problems and we begin to perceive these predicaments as opportunities. The outcome can be a more peaceful world, a diverse and vibrant world with new levels of expectation, and importantly, a world where hope is alive.

The world will unite in cyberspace before it does on the ground. From an education viewpoint, one of the evolving but ultimately most enabling artifacts of the knowledge age will be... cyberschools.

— Glenn R. Jones, 2010

APPENDIX ONE

Assessing Student Learning in the Knowledge Age:
The Jones International University Assessment Model

Joyce A. Scott, PhD and Robert W. Fulton, PhD

OVERVIEW

In 1983, the study "A Nation at Risk" called for reform in K–12 education but prompted a reaction in the higher education community as well (U.S. Department of Education, 1983). A distinguished Study Group on the Conditions of Excellence in American Higher Education responded in 1984 with a study on "Involvement in Learning: Realizing the Potential of American Higher Education" (U.S. Department of Education, 1984). It launched a conversation on assessment that has engaged higher education for the ensuing 25-years.

The Study Group fundamentally rejected the previous methods of defining educational excellence and advocated a shift from indicators based on inputs to those based on outcomes and, specifically, student learning outcomes (U.S. Department of Education, 1984, p.16). The Group called upon institutions to show the improvements in students' knowledge, skills and abilities between admission and graduation, to publish standards of performance for the degree and to demonstrate the efficiency and cost-effectiveness of their reforms. "The third condition of excellence is regular and periodic assessment and feedback. The use of assessment information to redirect effort is an essential ingredient in effective learning and serves as a powerful lever for involvement" (p.21). If the criteria for excellence were an important contribution, even greater was the insistence that students needed to be involved with their learning.

Similar themes were reprised in a report from the Association of American Colleges (AAC) the following year. Integrity in the College Curriculum: A Report to the Academic Community (Association of American Colleges, 1985) raised the central questions: "Why have colleges and universities failed to develop systems for evaluating the effectiveness of courses and programs? Why are they so reluctant to employ rigorous examining procedures as their students progress toward their degrees" (p.3)? Describing assessment as "an organic part of learning," the Select Committee responsible for the report called upon educators to evaluate students, programs and faculty regularly (p.13); to examine the curriculum constantly (p.12); and to track student progress in acquiring requisite abilities for their degrees, not only for the benefit of students but for the use of faculty in refining their own effectiveness (p.33).

The same year, the now-defunct American Association of Higher Education (AAHE) launched its first forum on assessment in partnership with the National Institute of Education. This opened the way to a series of conferences involving faculty from across the nation in a rich and extensive dialogue about assessment and how best to use it to improve their students' learning and their own teaching. Ewell (2002) traced the early history of assessment and public officials' concurrent demands for greater accountability from higher education as leading to requirements for reporting on learning outcomes in about half the states by 1989 (p.8). The growing public interest was reflected in new, 1989 regulations of the Higher Education Act directing accreditors to ensure that institutions were attending to student learning outcomes.

From the mid-1980s, the dialogue on assessment continued to evolve, supported by the conferences and publications of the AAHE, the Carnegie Commission for the Advancement of Teaching, the Pew Charitable Trusts, the regular dissemination of information through *Change* and disciplinary associations as well as the work of the Council for Higher Education Accreditation (CHEA) and the regional and specialized accreditors. In 1996, the AAHE published its "Nine Principles of Good Practice for Assessing Student Learning," summarizing much of the discussion. During the same period, the number of states requiring some form of reporting on student outcomes or using performance funding incentives increased. In response to the growing call for information, CHEA adopted its "Statement of Mutual Responsibilities for Student Learning Outcomes: Accreditation, Institutions and Programs" (CHEA, 2003).

In the 2000s, the Association of American Colleges and Universities (AAC&U, formerly AAC) convened a national panel to articulate a vision for a new academy in which assessment of student learning figured prominently. Their report, Greater Expectations: A New Vision for Learning as a Nation Goes to College (AAC&U, 2002), set forth principles of good practice as well as desirable learning outcomes in undergraduate education. It laid the groundwork for an extended AAC&U initiative, Liberal Education and America's Promise (LEAP), which has contributed significantly to the scholarship on student learning and assessment through research reports, articles, conferences and policy statements with other associations.

In September 2005, Secretary of Education Margaret Spellings charged a Commission on the Future of Higher Education to design a national strategy for postsecondary education to accommodate the needs of the nation's diverse population, the economy and the workplace. The Commission's report, "A Test of Leadership: Charting the Future of U.S. Higher Education" (U.S. Department of Education, 2006), noted shortcomings across higher education. Accreditation, for example, was judged deficient in both process and findings: "The growing public demand for increased accountability, quality and transparency coupled with the changing structure and globalization of higher education requires a transformation of accreditation" (p.14). In particular, the Commission recommended that accreditation stress student learning outcomes and degree completion over inputs and processes.

The Spellings Commission touched on many issues that the academic community had been deliberating since the 1980s: purposes of assessment, strategies for evaluating student performance, and diminishing public confidence in the higher education system. With regard to purpose, the Commission emphasized public accountability; as to strategy, it promoted nationally available standardized tests such as the Collegiate Learning Assessment or the Measure of Academic Proficiency and Progress (U.S. Department of Education, 2006, p.23); and to address public confidence, the Commission recommended a "consumer-friendly information database" to include data on institutions' performance and costs (p.20).

The Commission emphasized employers' dissatisfaction with college graduates' skills, a concern noted in AAC's 1985 report and subsequently in many other venues. Other studies, including "Liberal Education for the Twenty-first Century: Business Expectations" (Jones, 2005), highlighted graduates' deficiencies (p.34) and argued for continuous calibration of the learning standards and the curriculum to the needs of the larger society (p.37). Among the competencies sought by business leaders, Jones cited "basic academics (writing, math, science, technology, and global integration); application skills (integrated and applied learning, critical thinking); and soft skills (teamwork, ethics, diversity, and lifelong learning preparation)" (p.35).

Research conducted by Peter D. Hart Research Associates, Inc. for AAC&U revealed that business executives find graduates deficient in job readiness, specifically lacking in work ethic, commitment to the job, communication skills and ability to work with others (2006a, p.1). More recently, Peter D. Hart Research Associates interviewed a sample of 301 employers and reported results in the study: "How Should Colleges Assess and Improve Student Learning? Employers' Views of the Accountability Challenge" (2008). Employers preferred real-world assessments, which showed application of learning, and considered multiple-choice assessments and college transcripts unhelpful (p.1). Employers favored faculty evaluations of internships and community-based work followed by essays and portfolios to help them evaluate graduates' readiness for the workplace (p.7).

The above summary does not begin to reflect the wealth and complexity of thinking on assessment of student learning that has occurred over the preceding 25-years. Notwithstanding this extensive conversation and the common threads running through it, however, assessment has not taken firm hold in the academy. Ewell noted that the assessment "movement" has had an atypical fortune, having neither disappeared after a brief time nor become fully institutionalized after an extended period (2002, p.21). At many institutions, assessment remains a marginal activity, pursued because required rather than integrated into the fabric of courses and programs.

Such is not the case at Jones International University (JIU) where assessment is fundamental to institutional infrastructure and operations. As an emerging learning organization, JIU had the flexibility to create learner support and organizational planning systems drawing upon the recent scholarship of learning and assessment. JIU's first interest lies in ensuring students a substantive and meaningful learning experience. This means monitoring all aspects of program quality through student learning outcomes. Thus, the JIU assessment paradigm was designed incrementally over a decade and toward that end, drawing upon the wisdom of the academy and responding to the needs of students, faculty and the larger society, thereby creating a culture of evidence.

THE JIU ASSESSMENT PARADIGM

Creation of the JIU assessment system engaged professors, administrators, instructional designers, external advisory board members, including employers, and necessitated careful integration of standards set by professional associations and state licensing agencies. Development was indigenous, guided primarily by the needs of learners, employers and the institution (for purposes of accountability and continuous improvement).

With respect to students, JIU serves a highly diverse population largely comprised of working adults and military personnel, the majority of whom enroll part time. In fall 2008, there were 1882 students enrolled, with an average age of 36. Of these, 67 percent were women, 25 percent were African American and five percent were Hispanic. Fifty-four percent of graduate students were enrolled in business programs and 46 percent in education. To facilitate these students' learning, JIU organized its courses on an eight week format and offers weekly feedback on learning achievement, thereby building confidence and motivation.

Recognizing that many students pursue additional education to qualify for professional advancement, JIU designed programs and learning outcomes to reflect the standards of the relevant professions. Advisory boards involving practitioners and experts in a given field contributed the "big picture" perspective and identified the needs of their professions to ground mission and vision. Program committees, comprised of lead professors within a program and administrators, worked within that framework to develop specific courses and curricula, to map how courses meet program and institutional objectives, and to address academic issues. In 2006 and 2008, the

School of Business convened advisory boards to review all programs. These reviews resulted in two important adjustments to programs: 1) greater emphasis on ethical decision making and global citizenship in program outcomes and 2) modification of program outcomes to make them more actionable and measurable.

Where accrediting or professional associations had published formal standards to guide curriculum development and learning outcomes in a given program area, JIU adopted these criteria in formulating its programs. For example, the undergraduate and graduate accounting programs were configured to the standard expectations published by the American Institute of Certified Public Accountants, Inc.; and the program in finance was built around criteria from and ultimately approved by the American Academy of Financial Management™. Similar alignments occurred in education programs where National Council for Accreditation of Teacher Education (NCATE) and Colorado Department of Education standards guided curriculum development.

In response to corporate employers' long-standing criticism of college graduates' workplace skills, the Academic Committee, composed of professors, deans, instructional designers and other administrators, drew upon multiple sources to generate a set of institutional learning objectives or workplace competencies for all degrees. Each competency is accompanied by a succinct definition. For example, "teamwork" is characterized for the student by the following behaviors: "Cultivates inclusiveness, participation, team-based success, and mutual support. Respects diversity of opinion, belief, ethnic/cultural background. Is reliable and personally accountable to the team. Keeps commitments. [If a group project]." Feedback on relevant workplace competencies is incorporated into all course outcomes reports (see Table 1 on page 185).

The growing body of scholarship on assessment suggested criteria to guide formulation of the total assessment plan. Specifically, it should:

- be based on measurement of student learning outcomes (U.S. Department of Education, 1984, p.16);

- be built around clearly stated learning objectives of courses and curriculum as well as those of the institution (U.S. Department of Education,1984, p.21);

- provide learners, faculty and administrators with continuous and cumulative feedback about student learning and progress within courses and across the curriculum (Shavelson, 2007, p.23);

- permit aggregation of learner outcomes data for institutional purposes (Miller and Leskes, 2005, p.12);

- be integrated with institutional structures for evaluating faculty performance, course and curriculum quality, and institutional effectiveness (Association of American Colleges and Universities, 2002, p.36); and

- result in evidence to guide continuous institutional improvement and to respond to multiple external accountability requirements (U.S. Department of Education, 2006, passim).

With these criteria in mind, JIU personnel developed the Assessment of Student Learning and Institutional Improvement Model which is beneficially described through the framework proposed by Miller and Leskes (2005). Consistent with their model, the JIU assessment paradigm comprises five components: individual student learning within courses, individual student learning across courses, course effectiveness, program effectiveness and institutional effectiveness.

Basic components of the system include documents defining expectations for learning.

- Course learning objectives, developed by program professors and chairs, define the learning outcomes to be achieved in each course.

- Program learning objectives, similarly developed, set forth the learning expectations for the entire program and are published on the University's Web site:

  School of Education
  http://www.jonesinternational.edu/schools/education/med/ and

School of Business
http://www.jonesinternational.edu/schools/business/index.php

- Institutional learning objectives or workplace competencies outline skills and dispositions judged essential for effective participation in the 21st century workplace. These number 19 in all and were defined by the Academic Committee through research on employers' needs and consultation with corporate advisors. As appropriate, these competencies are selected for each course and evaluated along with the course learning goals.

Several assessment tools have been developed to support the system and ensure regular feedback, both to the student and the institution. Among these tools are the in-course assessment tool (scoring rubric), project sponsor appraisal tool, end-of-program assessment, satisfactory academic progress audit, graduation degree audit, professors' end-of-course survey, students' end-of-course survey, graduates' end-of-program survey, employers' end-of-program survey and professor self- and peer-appraisal tools. Recently completed is a comprehensive Student Success Data Report, which provides the student with an integrated and progressive analysis of his/her performance individually and in relation to others in the program.

Level 1: Assessing Individual Student Learning. Program committees involving professors and department chairs as well as advisory board members, where suitable, are responsible for setting program and course learning objectives which are subject to review. A broadly-representative Academic Committee, in consultation with program advisory boards, established all institutional learning objectives (workplace competencies), standards and processes; program vision, mission and professional standards; admission processes; and academic processes that guide student participation.

Transparency is a key to student success. At the beginning of every course, professors provide students with a detailed description of course expectations, including learning objectives, and an assessment tool (scoring rubric) that is used consistently within and across courses. Students are expected to monitor their own learning and are accountable for producing evidence of learning to meet the expectations. Using the same scale, professors evaluate every student and make students aware of the norms against which their performance is reviewed:

- Basic: The student demonstrates an understanding, but is not yet able to apply the learning outcome in the field;

- Developing: The student demonstrates increasing understanding and begins to apply the learning outcome in the field with assistance;

- Proficient: The student demonstrates a solid understanding and is able to apply the learning outcome in the field without assistance; and

- Advanced: The student demonstrates exemplary performance and skillful application of the learning outcome in the field.

The four ratings above are used to assess student learning of disciplinary outcomes and workplace competencies. To assess individual student learning, a professor provides formative feedback for each successive course module via the in-course assessment tool (scoring rubric) and rates the student's performance on all pertinent learning objectives. This is intended to help the student understand his/her progress in the course and to improve learning. However, the assessment tool also contains text boxes for the professor's final, detailed, qualitative appraisal of student performance. This summative analysis recounts the student's strengths and shortcomings (if any), raises questions, recommends how to improve and records the final grade. The final assessment is filed in the JIU data repository and made available to the student. This assessment provides the student information that grades alone cannot communicate and sets a direction for future growth.

Central to course design and learning assessment is the professional synthesizing project (PSP), an eight-week, course-embedded applied research project with multiple points of formative assessment that yields the primary evidence of student learning. The PSP is a vehicle by which students demonstrate their ability to apply theory and content learned in a course to the solution of a practical professional problem. Students demonstrate via their PSP that they have learned and can apply the knowledge, skills and dispositions necessary to succeed in their fields.

For advanced students, the PSP is in service to a sponsor in a professional field. Sponsored project-based learning rests on research that learning in adulthood is social in nature and firmly embedded in the life context of the learner (Knowles, Holton, & Swanson, 1998). Learning takes place in social groups engaged in common practice. In a community of practice, learning, practice and identity development are intertwined. Through sponsored projects, students connect to leaders in their professions and complete professionally meaningful work that changes their worlds for the better. Students open avenues for their personal career development and become professionals.

The PSP is structured to promote real-world, active learning in that it:

- Is drawn from the real needs of a given professional community and addresses a significant field-based question or problem;
- Engages a sponsor — a leader or leadership team in the field — who has a real need for the project to be conducted;
- Includes a step-by-step plan and a realistic timeline for completion within a course;
- Employs well-defined success measures, benchmarks, tasks, roles and responsibilities, resources and strategies; and
- Incorporates applicable course, program and institutional learning objectives in a synthesized and interdisciplinary manner.

Is assessed against learning objectives rooted in institutional, state and accreditation standards.

To be successful, a student project must exhibit the following features:

- Be thoroughly researched and include rich data from multiple sources with an analysis that is comprehensive and convincing;

- Demonstrate mature critical thinking as well as a thoughtful understanding of the literature and theory in the field;

- Develop an argument that is focused, logical, rigorous and sustained; and

- Result in a well-written report, with a point of view and confident voice, according to American Psychological Association (APA) Publication Manual standards.

Throughout the PSPs, professors stay in close communication with their students to lend support and offer guidance. In this process, as elsewhere, professors use the analytical scoring rubric to provide students with timely, detailed formative and summative, qualitative and quantitative project assessments. The final course grade is a holistic assessment of the student's evidence of learning in the course, but the grade is bolstered by substantial documentation of student performance. In addition to the traditional grade, the student receives reports on course outcomes and on workplace competencies, assessing his/her performance using the evaluative scale noted earlier. Table 1 reflects the supplementary evaluative information a student receives about performance on course objectives.

Table I. Student Course Outcomes Report

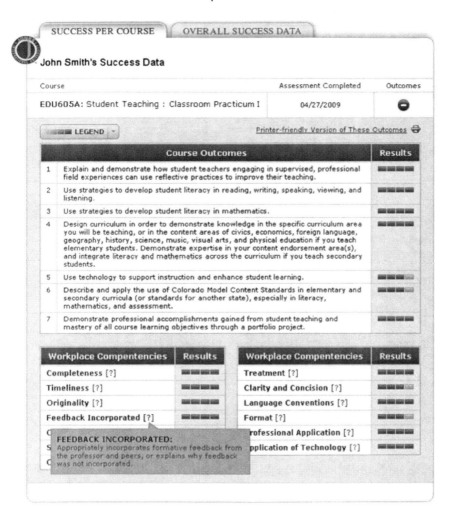

Table 2 summarizes the activities and artifacts involved in assessing individual student learning.

Table 2. Assessing Individual Student Learning within JIU Courses

Assessment	Source of Evidence	Assessor	Assessment Tool	Marker
Student Achievement of Course Learning Objectives	Professional Synthesizing Projects	Professors	In-Course Assessment Tool	In Each Course
Student Achievement of Institutional Learning Objectives (Workplace Competencies)	Professional Synthesizing Projects	Professors	In-Course Assessment Tool	In Each Course
Student Achievement of Specialty Professional Association (SPA) Learning Objectives	Professional Synthesizing Projects	Professors	In-Course Assessment Tool	In Each Course
Student Service to Sponsor	Professional Synthesizing Projects	Sponsors	In-Course Sponsor Appraisal Tool	In Each Course

Level 2: Assessing Individual Student Learning across JIU Courses.
In addition to being responsible for producing evidence of learning in
every course, students must show evidence of learning across courses.
Students gather evidence of their learning and integrate it over time.
This means that professors and instructional designers must work
together to ensure that there are clear learning objectives and aligned
assessment tools so that students can monitor themselves. Professors
are charged with assessing students' work objectively and providing
them holistic and analytic feedback.

To assess student learning across courses, professors provide each
student with feedback on every embedded end-of-course project and
a final summative assessment of his/her performance on all learning
objectives (Figure 1. Student Success Data from All Courses). These
data are progressively aggregated for each student on all learning
outcomes over the period of enrollment in a program, thus providing
meaningful, cumulative feedback (Figure 2. All Course Outcomes
by Quarter). This method allows the student and the institution to
track the degree to which s/he has improved during the program
and has achieved all pertinent learning objectives. The Student
Success Data Report is updated monthly and allows every student
to view an aggregation of his/her learning. Students can track their
personal development over time and strategize about ways to improve
their learning. Furthermore, the report is available to professors
and academic service counselors to ensure that every student gets
the support s/he needs to demonstrate achievement of all required
learning objectives. Whenever a student's work is found not to have
met course or institutional objectives, the professor is encouraged to
return the work ungraded with detailed feedback, supply samples of
exemplary work as a model, offer tutoring resources and provide an
opportunity for resubmission prior to assigning a final grade. The
University also recommends supplementary resources and online
tutoring services.

Figure 1. Student Success Data from All Courses

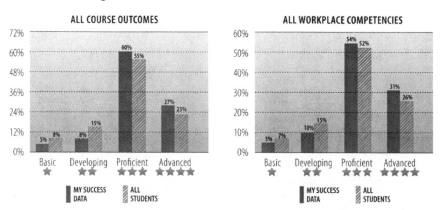

Figure 2. All Course Outcomes by Quarter

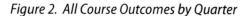

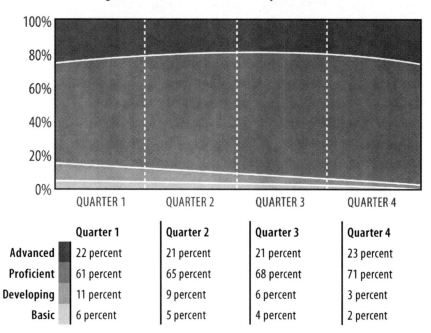

	Quarter 1	Quarter 2	Quarter 3	Quarter 4
Advanced	22 percent	21 percent	21 percent	23 percent
Proficient	61 percent	65 percent	68 percent	71 percent
Developing	11 percent	9 percent	6 percent	3 percent
Basic	6 percent	5 percent	4 percent	2 percent

The primary evidence of student learning across courses resides in the student's end-of-program professional synthesizing project (EPSP). The project may take many forms depending on the particular field of study: a capstone project, a portfolio of work samples, an internship, student teaching or a dissertation, for example. The course professor evaluates the EPSP, and it must be scored at least proficient on every program learning outcome for the student to graduate. In other words, the student must demonstrate a solid understanding of the discipline's content and apply the learning outcomes in the field without assistance.

JIU recognizes that grades provide holistic feedback but are difficult to interpret across courses; nevertheless, deans and the registrar track and report students' satisfactory academic progress (SAP) throughout each student's tenure and provide a comprehensive graduation degree audit at the end of the student's program. Doing so ensures that graduates complete all required courses and maintain the minimum required GPA. Like most colleges and universities, JIU uses course grades as a measure of students' academic achievement. Though grading is course specific, every effort is made to establish uniform program grading policies and procedures. As students pursue their degrees, they know how to maintain satisfactory academic progress and how to achieve academic excellence. When a student earns a grade in a course, the University is confident that the grade earned reflects the designated level of achievement of course and institutional learning objectives. Table 3 summarizes the activities and artifacts involved in assessing student learning across courses.

Table 3. Assessing Individual Student Learning across JIU Courses

Assessment	Source of Evidence	Assessor	Assessment Tool	Marker
Student Achievement of Program Learning Objectives	End-of-Program Professional Synthesizing Projects	Professors	End-of-Program Assessment Tool	End of Program
Student Achievement of Institutional Learning Objectives (Workplace Competencies)	End-of-Program Professional Synthesizing Projects	Professors	End-of-Program Assessment Tool	End of Program
Student Achievement of Specialty Professional Association (SPA) Learning Objectives	End-of-Program Professional Synthesizing Projects	Professors	End-of-Program Assessment Tool	End of Program
Student Satisfactory Academic Progress Course	Completion and GPA	Deans and Registrar	Satisfactory Academic Progress Audit	Monthly
Student Completion of Program	Course Completion and GPA	Deans and Registrar	Graduation Degree Audit	End of Program

Level 3: Assessing JIU Courses. The JIU Course Design Model involves broad participation from across the institution, as does the course assessment process. Course assessment is based on multiple sources of evidence: student demonstrations of course, institutional and professional learning outcomes; congruence of course to the course design model; professors' and students' end-of-course survey data; the professional development coordinator's appraisal of professors' performance; professors' self- and peer-appraisals; and the faculty administrator's analysis of relevant data. These assessments include an analysis of students' embedded assignments; course quality, consistency and cohesion; and instructional quality, consistency and cohesion.

The JIU course design model — adhering to best practices in instructional design — was conceived to ensure that all students in all courses demonstrate achievement of explicit course, institutional and program learning objectives. In addition, the model is intended to ensure that: the course level is appropriately targeted for students' abilities when they begin it; every student is well prepared for subsequent courses; different sections of the same course achieve similar learning outcomes; and any given course fulfills its purpose in the larger curriculum. Every course follows a rigorous model to ensure consistent formatting for course elements such as home page, syllabus, resources, announcements, workspace, profiles, grade book, library and supplementary resources or help.

Guidelines for course-level assessment are extensive and require professors to provide formative feedback on each student submission. Additional feedback is elicited, tracked and reported continuously to help professors improve their instruction and students' learning. At the end of every course, professors respond to an end-of-course survey to assess the class of students — collectively — on their achievement of the course learning objectives (see Table 4).

Table 4. JIU End-Of-Course Professor Survey

Course

1. This course had high standards that are appropriate for my profession.

2. Students' work in this course was high quality, and they learned what they were expected to learn.

3. This course allowed enough time for students to be successful.

4. Course materials and readings were current, important to the field and aided students' learning.

5. This course prepared students for later courses in the curriculum.

6. This course was appropriately targeted for students' abilities.

7. This course fulfills its purpose in the larger curriculum.

8. In this course, students demonstrated their ability to think and act like global citizens with personal integrity and ethical behavior.

9. In this course, students demonstrated their ability to think and act like leaders and supportive teammates.

10. In this course, students demonstrated their ability to adapt and apply theories to solve "real world" problems using innovative and creative solutions.

11. In this course, students demonstrated their ability to communicate effectively.

12. In this course, students demonstrated their ability to think analytically, critically and systemically to prepare for a life of learning and accomplishment.

13. In this course, students demonstrated their ability to leverage information and technology to amplify personal and organizational achievement.

Qualitative Answers

1. What did you like most about this course?

2. How would you improve this course?

3. What else would you like JIU to know about your experience in this course?

4. hich of the course objectives were most strongly accomplished by your students (please explain)?

5. Which of the course objectives were not or only weakly accomplished by your students (please explain)?

6. What is working best for the students in this course (please explain)?

7. What other suggestions do you have for this course (please explain)?

In like manner, students answer an end-of-course survey to assess the quality of their courses and their professors' performance (see Table 5).

Table 5. JIU End-Of-Course Student Survey

Overall Course Rating

1. Would you recommend this course to others?

2. Would you recommend this professor to others?

3. Would you recommend JIU to others?

Course

1. This course had high standards that are appropriate for my profession.

2. My work in this course was high quality, and I learned what I was expected to learn.

3. This course allowed enough time for me to be successful.

4. Course materials and readings were current, important to the field and aided my learning.

5. In this course, I demonstrated my ability to think and act like a global citizen with personal integrity and ethical behavior.

6. In this course, I demonstrated my ability to think and act like a leader and supportive teammate.

7. In this course, I demonstrated my ability to adapt and apply theories to solve "real world" problems using innovative and creative solutions.

8. In this course, I demonstrated my ability to communicate effectively.

9. In this course, I demonstrated my ability to think analytically, critically and systemically to prepare for a life of learning and accomplishment.

10. In this course, I demonstrated my ability to leverage information and technology to amplify personal and organizational achievement.

Professor

1. This professor had high expectations and challenged me to achieve.

2. This professor encouraged me to communicate with him/her.

3. This professor encouraged me to communicate with other students.

4. This professor knew the subject matter and was a great resource for learning.

5. This professor explained expectations at the start of the course.

6. This professor provided high quality and prompt feedback and grades.

7. This professor led class discussions to help me critically think about course content.

8. This professor valued and sought the active learning of all students.

Qualitative Answers

1. What did you like most about this course?

2. How would you improve this course?

3. What else would you like JIU to know about your experience in this course?

In addition to these data, professors and instructional designers review student work samples—primarily the embedded assessments from the PSP and EPSP—on a two year revision cycle to ensure high-quality, current and relevant courses. Every course is taught in multiple sections, so professor effectiveness can be measured via evidence from common course assignments.

As part of its continuous improvement strategy, JIU evaluates professors annually and asks them to complete a self-appraisal on teaching actions that led to student success or satisfaction. In addition, each professor is asked to appraise a peer's teaching on the same factors. The appraisal process and outcomes are used in conjunction with students' end-of-course survey results to aid in professor development. Table 6 summarizes the activities and artifacts associated with course assessment.

Table 6. Assessing JIU Courses

Assessment	Source of Evidence	Assessor	Assessment Tool	Marker
Student Achievement of Course Learning Objectives	Professional Synthesizing Projects	Professors	In-Course Assessment Tool	In Each Course

Student Achievement of Institutional Learning Objectives (Workplace Competencies)	Professional Synthesizing Projects	Professors	In-Course Assessment Tool	In Each Course
Student Achievement of Specialty Professional Association (SPA) Learning Objectives	Professional Synthesizing Projects	Professors	In-Course Assessment Tool	In Each Course
Alignment to Course Design Model	Courses	Instructional Designers	Course Features Report	Two-Year Revision Cycle
Aggregate Student Achievement of Course Learning Objectives	Professors' End-of-Course Survey Data	Professors	Professors' End-of-Course Survey	Monthly
Professors' Performance	Students' End-of-Course Survey Data	Students	Students' End-of-Course Survey	Monthly
Course Quality	Students' End-of-Course Survey Data	Students	Students' End-of-Course Survey	Monthly
Professors' Performance	Faculty Appraisal Data	Faculty Professional Development Coordinator	Faculty Appraisal Tool	Annually
Professors' Performance	Self Appraisal Data	Professors	Self-Appraisal Tool	Annually
Professors' Performance	Peer Appraisal Data	Professors	Peer-Appraisal Tool	Annually
Professors' Performance	Faculty Analysis Data	Faculty Administrator	Faculty Analysis Report	Quarterly

Level 4: Assessing JIU Programs. In the process of program development, professional advisory boards worked with deans, chairs and professors to align program vision and mission with professional standards. Professors, deans and the chief academic officer are charged with ensuring quality by reviewing and approving curriculum and strategies for instruction and assessment; establishing explicit degree and course learning objectives; and incorporating research on teaching and learning as well as technological advances that may affect student learning positively.

Program committees involving professors, instructional designers and administrators set program standards and assessment tools. Thereafter, it is their responsibility to ensure that JIU's programs achieve their student learning goals, and they use multiple methods for this purpose, drawing upon student success data as a primary resource. These data offer direct evidence of student learning from end-of-course and end-of-program professional synthesizing projects. By this means, professors ensure that students demonstrate achievement of program, institutional and professional learning objectives within their courses. Professors and administrators have carefully identified and limited the important assessment points for sampling and analysis and consider end-point data to be the most revelatory about how well a program has achieved its goals for student learning.

A data repository holds the information about student learning outcomes in perpetuity. Administrators may track students' aggregate achievement and compare and contrast student learning over time in a variety of ways, including all students within a given course; all students across multiple sections of a course; all students linked to a given professor; all students across all courses within a given program; and all students across a degree level (e.g., associates, bachelors, etc.). Information about sub-categories in a program can be aggregated to the program level. A sampling of student work designated basic, developing, proficient or advanced documents JIU's expectations for and assessment of student learning. The data can be disaggregated as well, to reveal how sub-groups of students compare to others in the program. When an issue of reliability arises, the University addresses it through professional development activities for professors and staff.

JIU employs external evaluators to assess a sample of each degree program's EPSPs. The evaluator uses the same assessment tool and program objectives (see sample in Table 7 for a finance specialization) as the original course professor. The evaluator's results are cross-tabulated with the professor's scores to measure inter-rater reliability. Results include cross-tabulation descriptive statistics and Cohen's Kappa (Kundel & Polansky, 2003). The second reading by a respected and objective external evaluator ensures that the larger learning community can trust the assessment results of its professors.

Table 7. Proficiencies for a Finance Specialization

Proficiencies – Finance Specialization

Problem-solve issues resulting from analyses of corporate finances, as well as their control mechanisms, and planning policies/processes.

Analyze the abilities of corporations to manage foreign exchange risk, cash, and capital budgeting issues in a global environment.

Use the investment and portfolio management tools employed by professional money managers.

Critically understand the opportunities, difficulties, and problems associated with the most current finance theories and issues confronting global corporations.

Other sources of program information include the end-of-program Graduates' Survey with a self-reflection on the learning process and a comparable survey for employers, who are invited to give an evaluation of graduates' learning. For students required to pass state content exams for licensure in initial educator preparation programs, their scores are also entered into the data repository to be tracked and compared over time and against state and national norms.

Assessment activities play a vital role in quality assurance and continuous improvement of JIU's courses and programs. Through program assessment processes, which align with Miller and Leskes' (2005) fourth level of assessment, professors and administrators confirm program purposes; document that the program fulfills its purposes; ensure that courses contribute to stated goals; investigate how well a program's design resonates with its objectives; ensure that courses are organized to allow for cumulative learning; and explore the level at which students perform on all program, institutional and professional learning objectives. Table 8 summarizes the sources of data and evaluation tools used in program assessment.

Table 8. Assessing JIU Programs

Assessment	Source of Evidence	Assessor	Assessment Tool	Marker
Aggregate Student Achievement of Program Learning Objectives	Student Success Data	Deans, Program Chairs and Professors	Student Success Data Report	Monthly
Aggregate Student Achievement of Institutional Learning Objectives (Workplace Competencies)	Student Success Data	Deans, Program Chairs and Professors	Student Success Data Report	Monthly
Aggregate Student Achievement of Specialty Professional Association (SPA) Learning Objectives	Student Success Data	Deans, Program Chairs and Professors	Student Success Data Report	Monthly
Aggregate Student Achievement on Standardized Professional Exams	Standardized Professional Exam Scores	External Agencies	Standardized Professional Exams	Quarterly & Annually

Aggregate Student Achievement of Program Learning Objectives	End-of-Program Professional Synthesizing Projects	External Evaluator	Random Sampling to Check Inter-rater Reliability	Quarterly & Annually
Aggregate Student Achievement of Program Learning Objectives	End-of-Program Graduates' Survey Data	Graduates	End-of-Program Graduates' Survey	Annually
Aggregate Student Achievement of Program Learning Objectives	End-of-Program Employers' Survey Data	Employers	End-of-Program Employers' Survey	Annually

Level 5: Assessing JIU. The chief academic officer and the deans are primarily responsible for assessing the institution. They created the JIU Assessment of Student Learning and Institutional Improvement Model and are responsible for its stewardship. They are supported by the Academic Committee which ensures full implementation of institutional assessment, using institutional data to improve teaching and learning and to align strategic planning and resource allocation accordingly. This robust assessment model cultivates student success and promotes continuous institutional renewal.

JIU uses multiple sources of evidence to appraise the University as a whole (Table 9). These include aggregate student success data; external evaluation of students' EPSPs; survey feedback from students, professors, graduates and employers; faculty analysis data; and aggregate academic metrics. These academic metrics comprise data on student recruitment and conversion rates, patterns of student enrollment and persistence, and degree completion rates.

Aggregate Student Success Data: At the end of a student's program, a professor provides him/her a detailed summative assessment of his/her performance on all pertinent learning objectives. With the scoring rubric, professors detail how well the final project demonstrates mastery of program and institutional learning objectives. This information is aggregated at the institutional level and used to guide revision of curriculum, instruction and assessment methods to ensure high-quality programs.

External Evaluation of Students' Projects: Independent, external evaluators review and summarize all assessment activities and student learning outcomes, using a sample of student work in end-of-program courses. The evaluator examines final projects and rates students' achievement for all learning objectives. In recent reviews of final projects, evaluators found that nearly all graduates performed at the proficient or advanced levels on critical learning objectives. This independent evaluation of student work serves two purposes: it validates the University's judgments about student achievement and its own effectiveness; and it alerts University officials to problems when they exist.

Surveys: JIU seeks feedback from students, professors, graduates and employers via surveys to assess the quality of both teaching and learning. Survey results from graduates and their employers feed into the continuous improvement process at all levels. Students' end-of-course surveys provide vital information about the quality of the curriculum, instruction, assessment and services. Professors evaluate their classes comprehensively and report whether or not course objectives were met, what works best in the course and what needs further development. Guided by Shulman's view that teaching is community property (2004, p.141), JIU publishes these survey results monthly and shares the report with professors and pertinent staff. All of these results go into a process of continuous improvement at all levels.

Faculty Analysis Data: Every professor is encouraged to achieve the University's faculty excellence award which involves a 360-degree evaluation for the preceding 12-months. The analysis of professors' professional performance and qualifications includes: annual appraisals (administrator, peer and self), scholarship (career and current), professional experience (including degree), teaching experience and student evaluations.

Aggregate Data Reports: JIU aggregates program and degree-level data to produce summary reports that are used for internal purposes to gauge how well student learning objectives are achieved. With few exceptions, assessment indices for the most recent reporting period confirm that JIU's graduates have achieved pertinent learning objectives. A model of the kind of aggregated performance data report the University is developing for all programs appears in Figures 3 and 4.

*Figure 3. Success Data from All Students in Bachelors Programs for 200_*

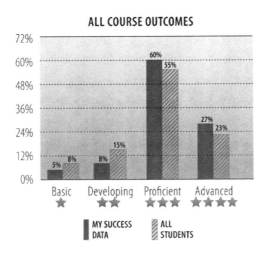

Figure 4. Workplace Competencies in Bachelors Programs

The academic leaders produce an annual assessment report. It summarizes evidence of student learning and upgrades to the assessment process and includes suggestions for JIU's future. These changes were introduced in the 2007-2008 reporting period:

- The whole-university model of assessment and institutional improvement described here and designed to showcase student achievement, drive program improvement and disseminate information about student learning outcomes quarterly;

- A data-driven scoring rubric for professors to assess student competence on course and institutional learning objectives, giving rise to an assessment data repository that provides documentation of learning for every student in every course;

- A revised course design model that includes formative and summative assessment of student learning and allows JIU to provide "evidence of student learning and teaching effectiveness that demonstrates it is fulfilling its educational mission" per the North Central Association's Higher Learning Commission Handbook of Accreditation (2003, p.3.1-4);

- A revised set of workplace competencies to be assessed in every course; and

- A database to track curriculum development and revision to support the University's goal of providing the most current and relevant curriculum available.

Table 9. Assessing JIU

Assessment	Source of Evidence	Assessor	Assessment Tool	Marker
Aggregate Student Achievement of Program Learning Objectives	Student Success Data	Deans, Program Chairs and Professors	Student Success Data Report	Monthly
Aggregate Student Achievement of Institutional Learning Objectives (Workplace Competencies)	Student Success Data	Deans, Program Chairs and Professors	Student Success Data Report	Monthly
Aggregate Student Achievement of Specialty Professional Association (SPA) Learning Objectives	Student Success Data	Deans, Program Chairs and Professors	Student Success Data Report	Monthly
Aggregate Student Achievement of Program Learning Objectives	End-of-Program Professional Synthesizing Projects	External Evaluator	Random Sampling to Check Inter-rater Reliability	Quarterly & Annually
Professors' Performance	Students' End-of-Course Survey Data	Students	Students' End-of-Course Survey	Monthly
Course Quality	Students' End-of-Course Survey Data	Students	Students' End-of-Course Survey	Monthly
Aggregate Student Achievement of Course Learning Objectives	Professors' End-of-Course Survey Data	Professors	Professors' End-of-Course Survey	Monthly

Aggregate Student Achievement of Program Learning Objectives	End-of-Program Graduates' Survey Data	Graduates	End-of-Program Graduates' Survey	Annually
Aggregate Student Achievement of Program Learning Objectives	End-of-Program Employers' Survey Data	Employers	End-of-Program Employers' Survey	Annually
Professors' Performance	Faculty Analysis Data	Faculty Administrator	Faculty Analysis Report	Quarterly
Aggregate Academic Metrics	Academic Metrics	Institutional Researcher	Academic Metrics Report	Monthly

INNOVATION IN THE SERVICE OF LEARNING

As public demands for greater accountability and transparency from higher education intensified in the 1990s, Ewell (1997) called for a new point of departure so the academy could answer its constituents. He judged that higher education had difficulty organizing for learning because the "system" was often constrained by traditional values, structures and modes of communication. In the long run, Ewell noted, "Results are what opinion leaders and the public understand best as the substance of accountability for higher education" (2008, p.136).

Jones International University offered this new point of departure. Founded in 1993, JIU lacked the above constraints and seized the opportunity to build a new "mental model" (Angelo, 1999) of what a university can be and do. Drawing upon the vision and expertise of its founder, professors, staff and administrators, the wealth of recent scholarship on learning and assessment, and the best practices of the academy, JIU created a new model where all educational practices were designed to promote student learning and to achieve the desired results. Organizing for learning meant that everyone affiliated with JIU became an intentional, involved, life-long learner (AAC&U, 2002; AAC&U, 2007). All continuously examine and reflect on the learning that occurs. Drawing upon these reflections and student data, colleagues deliberate and purposefully make changes to enhance the learning experience and the University's quality and effectiveness.

Intentional Learners: To foster intentionality, JIU focused the energies of the University on the student learning experience. Clear statements of course, program and workplace learning goals orient instruction and are used to measure student performance. Assessment tools and activities are embedded in courses and programs, reminding learners of the purpose of their endeavor (Shulman, 2007). Levels of achievement are defined as are the types of evidence that students must present in satisfaction of the goals (CHEA, 2003). Weekly and cumulative reports on achievement serve as valuable tools to corroborate progress (AAC, 1985, p.33), thereby validating student mastery and guiding future efforts. Student outcomes data, regularly shared within the broader JIU community, also inform instructional strategies, course design, program objectives and (re-)conceptualization. In biennial reviews, all of these factors are examined. Then, as needed, they are purposefully adjusted to enhance quality and to align with the changing needs of society, the professions and the workplace (AAC&U, 2007, p.2).

Involved Learners: Involvement for students means that they are expected to monitor their learning and reflect upon it continuously. The aim, here, is for them to develop self-awareness about how they learn and about how to apply their learning. Performance evaluations, course by course and over the extent of a program, allow students to track their progress individually and against aggregate peer data. Newly-minted, colorful student success reports give them a snapshot of their performance in any completed course and cumulatively. Students evaluate their courses and program regularly. Other types of student involvement may vary by program, but all students participate in discussion groups, real-world projects and capstone experiences favored by employers (Ewell, 2008; Peter D. Hart Research Associates, 2006 a and b; and 2008). Through their projects, students engage representatives from the professional community to serve as sponsors. Sponsors mentor and further involve students in the real-world aspects of their fields of study, helping them bridge the theory-practice gap (Gilroy, 2007, p.11).

Professors, trained extensively in evaluation techniques, give students weekly individualized feedback, thereby promoting their continuing engagement (Brookfield, 2006, p.193). At the end of their courses, professors submit individual narrative evaluations in addition to grades. The narrative, especially valued by employers (Peter D. Hart Research Associates, 2008), analyzes student progress and workplace "readiness." JIU encourages professors' involvement via many policies, including a recommended 24-hour response time to student inquiries. According to Brookfield, "Instructional responsiveness is central to the creation of an effective online learning environment" (2006, p.198). In addition to reflecting on student achievement, at the end of every course professors reflect on their own teaching and the quality of learning that they facilitated. This survey is collected and analyzed by University officers. Further, once a year a colleague "sits in" to assess another's teaching and contribute feedback for the course evaluation. Professors, staff and administrators all work collaboratively on course design, review and revision as well as on updating course and program objectives and workplace competencies. Finally, employers and graduates are involved through surveys about program currency, quality and results.

Life-long Learners. In Greater Expectations (2002), AAC&U envisioned a New Academy where educational practices are attuned to the needs of the 21st Century. The New Academy would be built on a culture of evidence: "In terms of its operations, the institution itself becomes a life-long learner, continuously evaluating and assessing itself at all levels, then feeding the results back into improvement loops for both student learning and campus processes" (2002, p.36). The previous account of the Assessment of Student Learning and Institutional Improvement Model demonstrates that JIU meets the standard. Members of the University community regularly consider how to advance the learning process: For example, one staff member suggested that individual and aggregate student success profiles — a form of baseline data — be generated for professors at the beginning of each course to assist them in adapting instruction and feedback to their individual and diverse learners.

If the prior summary of JIU's assessment model is comprehensive, it still does not address directly one additional component of the operation, the role of the JIU technological infrastructure. As technology resources and Internet access expanded in the 1990s, a discussion developed around their potential applications to higher education (Chickering & Ehrmann, 1996; Tyree, 1997) and continues today. Most of the benefits envisioned have been realized: the JIU technological infrastructure has facilitated frequent and responsive interaction between student and teacher, permitted more timely feedback, encouraged student-to-student interactions, documented student learning, and supported collection, analysis and dissemination of data such as surveys, student success reports and institutional research.

The infrastructure has taken learning to the students and, simultaneously, extended their reach to resources such as Jones e-global library®. Most importantly, however, the technological environment has promoted and sustained the progressive evolution of JIU as an authentic learning organization. Newman, Couturier and Scurry remarked that an assessment system sufficient to higher education's needs, both internal and external, could not be simplistic or cumbersome if it were to be practiced every day (2004, p.140). Although neither simplistic nor cumbersome, the JIU assessment model is complex and involves many participants. Such a learning tool would not have been manageable without sophisticated technological support and regular adaptation to its needs. Indeed, the absence of such support may account for why some institutions have been slow to embrace comprehensive learning assessment.

JIU's assessment model emerged from its academic community and developed indigenously, lending it integrity and credibility. It has grown and evolved with the institution and continues to do so. Today, as policymakers and accreditors seek greater accountability around student learning, new providers of assessment services have surfaced (Hutchings, 2009) offering campuses help with a variety of assessment tasks. Hutchings and Ewell both caution that a "quick fix" for administrative reporting purposes may not be sufficient to engage faculty to deal with the substance of learning (Hutchings, 2009; Ewell, 1997). Whether retro-fitting institutions with these products can improve learning or not remains to be seen. Certainly without deep faculty involvement, it is unlikely.

Innovation: Innovation comes hard to venerable and conservative institutions such as universities. The higher education literature of the past 25-years is replete with calls for change in the academic enterprise. Similarly, recent debates on accreditation and the reauthorization of the Higher Education Opportunity Act of 2008 made it clear that the public wants change, in the form of greater accountability, especially for student learning. Yet, the system has found it difficult to respond, and regional accreditors are now bringing pressure to bear and offering training about how to develop the desired culture of evidence.

Jones International University came into being at a propitious time and organized itself around learning, as the best way to serve its intended student population. JIU established a climate where innovation and experimentation were encouraged, where cutting-edge technologies were turned to the service of instruction, where technical expertise was readily available, where adequate financial resources were invested and where group and interdisciplinary problem solving was cultivated. JIU's professors, mostly trained and experienced at conventional campuses, were not constrained by the traditions of a single discipline or institution but were free to create their own models of teaching, learning and assessment.

In this climate and with the expertise of its accreditors and advisors, JIU has designed an assessment model suited to the Knowledge Age. Through it, the University can document student achievement and show how well it is fulfilling its learning goals and mission. Further, Jones International University can monitor its effectiveness as a learning organization and pursue continuous improvement. From this endeavor has emerged an institution that more than meets AAC&U's criteria for the New Academy as well as the needs of 21st Century learning.

Assessing Student Learning in the Knowledge Age: The Jones International University Assessment Model

REFERENCES

American Association for Higher Education. (1996). Nine principles of good practice for assessing student learning. Washington, DC: Author. Retrieved July 1, 2009, from http://teaching.uncc.edu/resources/best-practice-articles/assessment-grading/9-principles

Angelo, T.A. (1999, May). Doing assessment as if learning matters most. AAHE Bulletin.

Retrieved July 1, 2009, from http://education.gsu.edu/ctl/outcomes/ Doing percent20Assessment percent20As percent20If percent20Learning percent20Matters percent20Most.htm

Association of American Colleges (AAC). (1985). Integrity in the college curriculum: A report to the academic community. Washington, D C: Author.

Association of American Colleges and Universities (AAC&U). (2002). Greater expectations: A new vision for learning as a nation goes to college. Washington, DC: Author.

Association of American Colleges and Universities (AAC&U). (2007). College learning for the new global century: A report from the national leadership council for liberal education and America's promise. Washington, DC: Author.

Brookfield, S. D. (2006). The skillful teacher: On technique, trust and responsiveness in the classroom, (2nd ed.) San Francisco: Jossey-Bass.

Chickering, A. & Ehrmann, S. (1996, October). Implementing the seven principles: Technology as a lever. AAHE Bulletin. Retrieved July 28, 2009, from http://aahea.org/bulletins/articles/sevenprinciples.htm

Council for Higher Education Accreditation (CHEA). (2003). Statement of mutual responsibilities for student learning outcomes: Accreditation, institutions and programs. Washington, DC: Author. Retrieved July 21, 2009, from http://www.chea.org/pdf/ StmntStudentLearningOutcomes9-03.pdf

Ewell, P. (1997, December). Organizing for learning: A new imperative. AAHE Bulletin. Retrieved July 20, 2009, from https://www.aahea.org/bulletins/articles/ewell.htm

Ewell, P. (2002). An emerging scholarship: A brief history of assessment. In T. Banta and Associates, Building a scholarship of assessment. (pp. 3-25). San Francisco: Jossey-Bass.

Ewell, P. (2008). U.S. Accreditation and the future of quality assurance: a tenth anniversary report from the Council for Higher Education Accreditation. Washington, DC: Council for Higher Education Accreditation.

Gilroy, M. (2007, August 27). Broader skill sets needed in global economy. Hispanic Outlook, 17, 10-12.

Hutchings, P. (2009, May/June). The new guys in assessment town. Change. Retrieved May 17, 2009,   from http://www.changemag.org/May-June percent202009/full-assessment-town.html

Jones, R. (2005, Spring). Liberal education for the twenty-first century: Business expectations. Liberal Education, 32-37.

Knowles, M., Holton, E., & Swanson, R. (1998). The adult learner: The definitive classic in adult education and human resource development (5th ed.). Houston: Gulf Publishing Co.

Kundel, H L. & Polansky, M. (2003, August). Measurement of observer agreement. Radiology, 228: 303-308.

Miller, R. & Leskes, A. (2005). Levels of assessment: From the student to the institution. Washington, DC: Association of American Colleges and Universities.

Newman, F., Couturier, L., & Scurry, J. (2004). The Future of higher education: Rhetoric, reality and the risks of the market. San Francisco: Jossey-Bass.

North Central Association Higher Learning Commission. (2003). Handbook of accreditation. Chicago: Author.

Peter D. Hart Research Associates, Inc. (2006a, January 31). Report of findings based on focus groups among business executives. Washington, DC: Association of American Colleges and Universities. Retrieved July 10, 2009, from http://www.aacu.org/leap/documents/PrivateEmployersFindings.pdf

Peter D. Hart Research Associates, Inc. (2006b, December 28). How should colleges prepare students to succeed in today's global economy? Washington, DC: Association of American Colleges and Universities. Retrieved July 10, 2009, from http://www.aacu.org/leap/documents/Re8097abcombined.pdf

Peter D. Hart Research Associates, Inc. (2008, January 9). How should colleges assess and improve student learning? Employers' views of the accountability challenge. Washington, DC: Association of American Colleges and Universities. Retrieved July 10, 2009, from http://www.aacu.org/leap/documents/2008_Business_Leader_Poll.pdf

Shavelson, R. (2007). A brief history of student learning assessment: How we got where we are and a proposal for where to go next. Washington, DC: Association of American Colleges and Universities.

Shulman, L. (2004). Teaching as community Property: Essays on higher education. San Francisco: Jossey-Bass.

Shulman, L. (2007, January/February). Counting and recounting: Assessment and the quest for accountability. Change, 20-25.

Tyree, T. (1997, October). Assessing with the net: Using technology to know more about students. AAHE Bulletin. Retrieved July 20, 2009, from http://aahea.org/bulletins/articles/assessnet.htm

US Department of Education. (2006). A test of leadership: Charting the future of U.S. higher education. Final report of the Secretary's Commission on the Future of Higher Education. Washington, DC. Retrieved July 28, 2009, from http://www.ed.gov/about/bdscomm/list/hiedfuture/reports/final-report.pdf

U.S. Department of Education. National Commission on Excellence in Education. (1983). A nation at risk: The imperative for education reform. . Washington, DC: Author.

U.S. Department of Education. National Institute of Education. Study Group on the Conditions of Excellence in American Higher Education. (1984). Involvement in learning: Realizing the potential of American higher education. Washington, D.C.: Author.

FOOTNOTES

1   I. Elaine Allen and Jeff Seaman, "Online Nation: Five Years of Growth in Online Learning," October 2007, Sloan-C, page 4.

2   National Center for Education Statistics, *Distance Education at Degree-Granting Postsecondary Institutions 2006-2007*, December 2008, U.S. Department of Education, page 1.

3   "Clash of the Clouds," *The Economist*, October 15, 2009, accessed from www.economist.com.

4   "E-Learning Glossary," at the American Society for Training and Development (ASTD) web site, accessed at http://www.astd.org/LC/glossary.htm

5   *Bartlett's Familiar Quotations*, by John Bartlett, (New York: Little, Brown and Company, 2002), page 72.

6   Chris Anderson, *The Long Tail*, (New York: Hyperion, 2006), page 5.

7   Don Tapscott, *Grown Up Digital*, (New York: McGraw-Hill Books, 2009), page 122.

8   "A World of Witnesses," from a "Special Report on Mobility," *The Economist*, April 10, 2008, page 11.

9   *Respectfully Quoted*, edited by Suzy Platt, (Washington, DC: The Library of Congress, 1989), page 97.

10 Will and Ariel Durant, *The Lessons of History*, (New York: Simon and Schuster, 1968), page 79.

11 "Key Global Telecom Indicators for the World Telecommunication Service Sector," International Telecommunication Union, accessed from www.itu.int.

12 John Battelle, *The Search: How Google and Its Rivals Rewrote the Rules of Business and Transformed our Culture*, (New York: Portfolio, 2005), page 280.

13 William Butler Yeats, *The Norton Anthology of English Literature*, (New York: W.W. Norton & Company, 1975), page 2361.

14 Mark Toner, "Next Generation E-Readers Become Reality This Year," PRESSTIME, published by the Newspaper Association of America, January 2009, available online at http://www.naa.org.

15 *The Digest of Education Statistics 2008*, (Washington, DC: National Center for Education Statistics, U.S. Department of Education, March 2009), page 578.

16 "Homo mobilis: As language goes, so does thought," within the special report "Nomads at last: A special report on mobile telecoms," in *The Economist*, April 12, 2008, page 13.

17 "Google founder dreams of Google implant in your brain: Body modification—or channel ploy?" by Andrew Orlowski, posted March 3, 2004, in *The Register*, accessed online from www.theregister.co.uk.

18 Peter F. Drucker, *Landmarks of Tomorrow* (New York: Harper & Row, 1959).

19 Robert B. Reich, *The Work of Nations: Preparing Ourselves for 21ˢᵗ Century Capitalism* (New York: Alfred A. Knopf, 1991), page 225.

20 Will and Ariel Durant, *Lessons of History*, page 101.

21 *Annual Report of the Librarian of Congress For the Fiscal Year Ending September 30, 2007*, The Library of Congress.

22 William B. Johnston and Arnold H. Packer, *Workforce 2000* (Indianapolis, Ind.: Hudson Institute, 1987), page xxvii.

23 Richard W. Judy and Carol D'Amico, *Workforce 2020*, (Indianapolis, Ind.: Hudson Institute, 1997), pages 121-122.

24 Chris Anderson, *The Long Tail*, (New York: Hyperion, 2006), page 70.

25 Lao Tsu quotation from the web site *www.creatingminds.org*.

26 Peter Drucker, *The Atlantic Monthly*, November, 1994, available at http://www.theatlantic.com/politics/ecbig/soctrans.htm

27 Daniel J. Boorstin, *The Republic of Technology: Reflections on our Future Community*, (New York: Harper & Row, Publishers, 1978), page 96.

28 *The Digest of Education Statistics 2008*, op. cit., page 577-578.

29 "University top 200 in full," *Times Online*, October 9, 2008, accessed from *www.timesonline.co.uk*.

30 Ibid.

31 The Digest of Education Statistics 2008, op. cit., page 581.

32 Ibid.

33 Ibid, page 1.

34 Ibid, page 1.

35 *Highlights from PISA, 2006: Performance of U.S. 15 Year-Old Students in Science and Mathematics Literacy in an International Context* , U.S. Department of Education, page iii.

36 *EFA Global Monitoring Report 2006, Education for All—Global Monitoring Report*, UNESCO, page 19.

37 Sarah Garland, "Beyond the Diploma Mills," *Newsweek*, posted online at *www.newsweek.com* on December 13, 2008 and from the magazine issue dated December 22, 2008.

38 Ibid.

39   *EFA Global Monitoring Report 2006: Education for All — Global Monitoring Report*, UNESCO, page 19-20.

40   *The Digest of Education Statistics 2007*, page 577.

41   *The Digest of Education Statistics 2008*, page 581.

42   Ibid.

43   Mary Pat Seurkamp, "Changing Student Demographics," in *University Business*, October 2007, accessed from www.universitybusiness.com.

44   For detailed discussions of the importance of lifelong learning to the U.S. economy, see James Botkin et al., *Global Stakes: The Future of High Technology in America* (Cambridge, Mass.: Ballinger Publishing Company, 1982); William B. Johnston and Arnold H. Packer, *Workforce 2000: Work and Workers for the Twenty-first Century* (Indianapolis, Ind.: Hudson Institute, 1987), xxvi–xxvii and 95–103; Jack E. Bowsher, *Educating America: Lessons Learned in the Nation's Corporations* (New York: John Wiley & Sons, Inc., 1989), 208–220; and *A Nation at Risk: The Full Account* (Cambridge, Mass.: USA Research, for The National Commission on Excellence in Education, 1984).

45   Ken Dychtwald, Tamara J. Erickson, Robert Morison, *Workforce Crisis*, (Watertown, Mass.: Harvard Business School Press, 2006), page 160.

46   Communication from the Commission to the Council, The European Parliament, The European Economic and Social Committee and the Committee of the Regions, "Action Plan on Adult learning: It is always a good time to learn," *http://ec.europa.eu/education/policies/adult/com558_en.pdf page 9/27/07, page 10.*

47   *The Digest of Education Statistics 2008*, page 581.

48   Clayton Christensen, Curtis W. Johnson, Michael B. Horn, *Disrupting Class: How Disruptive Innovation Will Change the Way the World Learns*, (New York: McGraw-Hill, 2008), page 98.

49   Fast Statistics about Online Learning, iNACOL, May 2008, page 1, accessed from *https://www.iNACOL.org/media/iNACOL_fast_facts.pdf*

50   Patti Shank, "Thinking Critically to Move e-Learning Forward," by, Learning Peaks, LLC, Chapter 1 in *The E-Learning Handbook: Past Promises, Present Challenges*, editors Saul Carliner and Patti Shank, Pfeiffer, a Wiley Imprint, 2008, page 16.

52   Global Education Digest 2007, The UNESCO Institute for Statistics, 2007, page 10.

53   *The Digest of Education Statistics 2008*, pages 582-583. Population figures are as of midyear 2006; student enrollment figures reflect enrollments during the 2005-2006 year.

54   *Workforce Crisis*, op. cit., page 10-11.

55   *Global Education Digest 2007*, UNESCO Institute for Statistics, page 3.

56   Ibid.

57   John M. Bridgeland, John J. DiIulio, Jr. and Karen Burke Morison, *Silent Epidemic: Perspectives of High School Dropouts*, A Report by Civic Enterprises in association with Peter D. Hart Research Associates for the Bill & Melinda Gates Foundation, March 2006, page i.

58   In the statistics in this section on global learning, the various forms of distance education are compared but have different definitions; distance education is a subset of online or e-learning and generally refers to a method of learning where, if a teacher is involved, that teacher is located remotely. Online or e-learning can involve face-to-face instruction and can be a supplement to face-to-face instruction. LMS's can be used in the context of both distance education and face-to-face learning.

59   Allison Powell and Susan Patrick, "An International Perspective of K–12 Online Learning: A Summary of the 2006 iNACOL International E-Learning Survey," November 2006, pages 10-11.

60   Ibid, pages 18-24.

61   *EFA Global Monitoring Report 2008: Education for All by 2015: Will We Make It?*, UNESCO Publishing, Oxford University Press, 2007, page 134.

62   Ibid.

63   Interview with Ruwan Salgado, Executive Director of World Links.

64   "E-Learning in Tertiary Education," OECD Policy Brief, December 2005, pages 2-3.

65   For an analysis of the changing ratio of older students to traditional students and of the effects of the change, see Arthur Levine and Associates, *Shaping Higher Education's Future: Demographic Realities and Opportunities, 1990–2000* (San Francisco, Jossey-Bass Publishers, 1989), and current issues of *The Chronicle of Higher Education*.

66   *The Digest of Education Statistics 2008*, page 280.

67   Ibid, page 269.

68   Ken Dychtwald and Joe Flower, *Age Wave: The Challenges and Opportunities of an Aging America* (Los Angeles: Jeremy P. Tarcher, Inc., 1989), page 3.

69   Nicholas Eberstadt, "Growing Old the Hard Way: China, Russia, India," *Policy Review*, April & May 2006, accessed from www.hoover.org/publications/policyreview

70   Ibid.

71   Nicholas Eberstadt, "Power and Population in Asia," *Policy Review*, page 26, February and March, 2004.

72   Ray Marshall and Marc Tucker, *Thinking for a Living: Education and the Wealth of Nations* (New York: Basic Books, a division of HarperCollins Publishers, Inc., 1992), pages 44 and 49.

73   Penelope Wang, "Is College Still Worth the Price?" *Money Magazine*, August 22, 2008, retrieved from www.cnnmoney.com

74   Will and Ariel Durant, *Lessons of History*, page 101.

75   *The Digest of Education Statistics 2008*, page 269.

76   *The Digest of Education Statistics 2007*, page 3.

77   Chester E. Finn and Bruno V. Manno, "What's Wrong With the American University," *Wilson Quarterly*, Winter 1996, pages 44–45.

78 "A Test of Leadership: Charting the Future of U.S. Higher Education," A Report of the Commission Appointed by Secretary of Education Margaret Spellings, September 2006, U.S. Department of Education, page 10, found at this web site: http://www.ed.gov

79 "Trends in College Pricing, 2007," *The College Board*, pages 12, 18, 19. www.thecollegeboard.com. The price includes tuition, fee, room and board (TFRB) and are at constant 2007 dollars.

80 *The Digest of Education Statistics, 2007*, pages 261-262. Latest statistics are from 2005.

81 Tamar Lewin, "State Colleges Also Face Cuts in Ambitions," *The New York Times*, March 16, 2009, accessed from www.nytimes.com.

82 Penelope Wang, "Is College Still Worth the Price," *Money Magazine*, August 22, 2008, accessed from http://money.cnn.com

83 Thomas L. Friedman, *The World is Flat*, (New York: Farrar, Strauss and Giroux, 2005), page 290.

84 *Disrupting Class: How Disruptive Innovation Will Change the Way the World Learns*, op. cit., page 101.

85 John Zogby, *The Way We'll Be: The Zogby Report on the Transformation of the American Dream*, (New York: Random House, 2008), page 82. For a fuller treatment of Zogby's views on higher education and online education, read pages 80-82. Zogby also cites some of the issues that online learning still needs to overcome, in particular the quality of learning one receives online versus at a traditional university.

86 *Education at a Glance 2007, OECD Indicators*, Organisation for Economic Co-Operation and Development, page 141. When referring in this example to "university" education, the author is referring to OECD's "tertiary-type A" education.

87 Ray Uhalde and Jeff Strohl with Zamira Simkins, "America in the Global Economy: A Background Paper for the New Commission on the Skills of the American Workforce," December 2006, page 12, National Center on Education and the Economy, http://www.skillscommission.org

88 *Occupational Outlook Quarterly* (Fall 2004), page 32, accessed from http://www.bls.gov/opub/ooq/2004/fall/oochart.htm

89 Bureau of Labor Statistics Economic News Release, December 4, 2007, "Employment Projections: 2006-16 Summary," accessed from http://www.bls.gov/news.release/ecopro.nr0.htm

90 Wilson P. Dizard, Jr., *The Coming Knowledge Age: An Overview of Technology, Economics, and Politics*, Third Edition (New York: Longman Inc., 1989), pages 97–105.

91 *Cyberschools*, 2002, page viii, from foreword by Alvin and Heidi Toffler..

92 "The Brains Business," *The Economist*, September 10, 2005, pages 3-4.

93 Tom Peters, *Thriving on Chaos: Handbook for a Management Revolution*, (New York: Harper & Row, 1987), page 5.

94 Gary Duncan, "Lehman Brothers collapse sends shockwaves round the world," Times Online, September 16, 2008, downloaded from http://business.timesonline.co.uk/tol/business/.

95 Ken Dychtwald, Tamara Erickson, and Bob Morison, "It's Time to Retire Retirement," *The Harvard Business Review*, March 2004, page 18.

96 Rainer Strack, Jens Baier and Anders Fahlander, "Managing Demographic Risk," *The Harvard Business Review*, February, 2008, page 5.

97 Ibid, page 4.

98 Ken Dychtwald, Tamara Erickson, and Bob Morison, "It's Time to Retire Retirement," *The Harvard Business Review*, March 2004, pages 19, 20.

99 Ibid, page 21.

100 Ibid.

101 Chester Finn, "Twenty Five Years Later: A Nation Still at Risk," *The Wall Street Journal*, April 26, 2008, page A7.

102 Ibid.

103 Richard M. Ingersoll, "Teaching Science in the 21st Century: The Science and Mathematics Teacher Shortage: Fact and Myth," NSTA Reports, May, 2007, page 7.

104 Forrest Broman, "Opportunities Galore for Good Teachers," *The International Educator*, December, 2007, page 1.

105 U.S. Department of Education Press Release, "U.S. Secretary of Education Margaret Spellings Delivers Keynote at National Math Panel Forum," October 7, 2008, http://www.ed.gov/news/pressreleases/2008/10/10072008.html.

106 Linda Darling-Hammond, "Evaluating 'No Child Left Behind," *The Nation*, May 21, 2007, retrieved from http://www.thenation.com.

107 *Education at a Glance OECD 2008*, page 237.

108 Amanda Ripley, "Rhee Tackles Classroom Challenge," *Time magazine*, November 26, 2008, from www.time.com.

109 U.S. Department of Education Press Release, "Secretary Arne Duncan Speaks at the National Science Teachers Association Conference," March 20, 2009.

110 *Disrupting Class: How Disruptive Innovation Will Change the Way the World Learns*, op. cit., page 98.

111 Chi-Sing Li and Beverly Irby, "An Overview of Online Education: Attractiveness, Benefits, Challenges, Concerns and Recommendations," *College Student Journal*, June, 2008, page 451.

112 Marshall McLuhan, *Understanding Media: The Extensions of Man*,(New York: New American Library, 1964),

113 Marshall McLuhan and Quentin Fiore, *The Medium is the Message: An Inventory of Effects*, (New York: Bantam Books, 1967), page 26.

114 *Open Universities Australia 2007 Annual Report*, page 2, downloaded from https://www.open.edu.au/wps/portal.

115 OpenCourseWare Consortium web site, www.ocwconsortium.org.

116 Information on "Contact North at a Glance," web site found at *www.contactnorth.ca*.

117 Insung Jung, "Quality Assurance Survey of Mega Universities," in *Perspectives on Distance Education: Lifelong Learning & Distance Higher Education*, UNESCO, 2005, page 80-81.

118 "The Historical Development of Distance Education through Technology," by Denise M. Casey, *TechTrends*, March/April 2008.

119 Michael G. Moore and Melody M. Thompson, *The Effects of Distance Learning: A Summary of Literature* (University Park, Pa.: The American Center for the Study of Distance Education, 1990), page 3.

120 D. N. Wood and D. G. Wylie, "Reaching New Students Through New Technologies," *Educational Telecommunications* (Belmont, Calif.: Wadsworth Publishing Company, 1977), page 33.

121 Starr Roxanne Hiltz, *The Virtual Classroom* (Norwood, N.J.: Ablex Publishing Corporation, 1994), page 4.

122 The Pew Internet and American Life Project, "The Internet Goes to College," September 15, 2002, page 2.

123 The Pew Internet & American Life Project, "Writing, Technology & Teens," April 24, 2008, page ii.

124 "The 2008 Campus Computing Survey," The Campus Computing Project, accessed from www.campuscomputing.net.

125 Pew Internet Project Data Memo, January 28, 2009, Pew Internet & American Life Project, pages 1-2.

126 In early 1999, the North Central Association of Colleges and Schools, one of the six key accrediting bodies for higher education in the U.S., granted accreditation to Jones International University (JIU). JIU, which my company founded and operates, was the first purely virtual higher education institution to achieve this status. In 2008, JIU was also named one of the top 10 online universities in the world by "The Best Worst in Online Degree Programs," Global Academy Online, 2008, page 38.

127 Chris Evans, *Computers & Education*, February 2008, pages 491-498,

128 "ACU first university in nation to provide iPhone or iPod touch to all incoming Freshmen," Abilene Christian University Press Release, February 27, 2008, accessed from the Web at http://www.acu.edu

129 The New Media Consortium and the EduCause Learning Initiative, *The Horizon Report: 2008 Edition*, 2008, page 3.

130 Starr Roxanne Hiltz, *The Virtual Classroom*, (Norwood, N.J.: Ablex Publishing Corporation, 1994), page 4-5.

131 Parker Rossman, *The Emerging Worldwide Electronic University: Knowledge Age Global Higher Education*, (Westport, Conn.: Greenwood Press, 1992), pages 46-47.

132 Email sent from Starr Roxanne Hiltz.

133 *The Emerging Worldwide Electronic University*, op. cit., page 6.

134 2009 U.S. Department of Education report, "Evaluation of Evidence-Based Practices in Online Learning — A Meta-Analysis and Review of Online Learning Studies" — page xiv

135 "The Future of Online Learning: A Threshold Forum," *Threshold*, A Cable in the Classroom Publication, Fall 2008, page 18,

136 John Watson, Evergreen Consulting Associates, "Online Learning: The National Landscape," *Threshold*, A Cable in the Classroom Publication, Fall 2008, page 5.

137 Bill Tucker, "Florida's Online Option," *Education next*, Summer 2009, page 13.

138 Ibid, page 25.

139 Western Governors' Association, *From Vision to Reality: A Western Virtual University* (Denver, Colo., 1996), page 1.

140 "The World Is Open" by Dr. Curtis J. Bonk, page 52

141 James H. Billington, Librarian of Congress, "Opening Remarks to the UNESCO-Library of Congress Experts Meeting on the World Digital Library," Paris, November 30, 2006, downloaded from http://www.worlddigitallibrary.org.

142 "Internet Archive and Sun Microsystems Create Living History of the Internet," Press Release from Sun Microsystems, March 25, 2009.

143 On December 14, 2004, Google announced that the Libraries of Harvard, Stanford, the University of Michigan, the University of Oxford, and The New York Public Library were joining with Google to digitally scan library books and make them searchable online. Publishers and authors filed two copyright lawsuits in 2005. Copyright laws are a sticky issue in the world of digitization and virtual electronic libraries. Currently, documents not considered to be in what is called the public domain are subject to copyright laws. Libraries and other institutions that want to put copyrighted material on the World Wide Web or other electronic bulletin board systems still must get permission from the materials' authors or their estates Google settled these lawsuits in 2008 for $125 million, according to a Google press release. The Google book settlement was undergoing revision at presstime and the lawsuit regarding YouTube was still pending.

144 Jeff Jarvis, *What Would Google Do*, New York: HarperCollins, 2009), page 122.

145 "Google Checks out Library Books," December 14, 2008, Press Release From Google, accessed from http://www.google.com/press/pressrel/print_library.html

146 Speech by Eric Schmidt at the Economic Club of Washington DC on June 9, 2008.

147 Ibid.

148 Q&A with Eric Schmidt by Ken Auletta of *The New Yorker* at a Newhouse School event, June 11, 2008, video downloaded from http://www.youtube.com

149 Jeff Jarvis, *What Would Google Do*, op. cit., page 211.

150 Starr Roxanne Hiltz, *Virtual Classroom*, op cit., page 20.

151 Interview with John Perry Barlow in "What Are We Doing Online?" *Harper's Magazine*, August 1995, page 36.

152 "K–12 Online Learning: A Survey of U.S. School District Administrators," by Anthony G. Picciano and Jeff Seaman, Sloan-C, page 7, downloaded from http://www.sloan-c.org

153 John W. Verity, "The Internet: How It Will Change the Way You Do Business," *Business Week*, 11/14/94,

154 Nikhil Hutheesing, "Web Snarl," *Forbes*, April 8 1996, page 100.

155 Chester E. Finn and Bruno Manno, "What's Wrong With the American University?" *Wilson Quarterly*, (Winter 1996), page 48.

156 Ibid.

157 Dan Corrigan, *The Internet University: College Courses by Computer* (Harwich, Mass.: Cape Software Press, 1996).

158 "Best Worst in Online Degree Programs," Global Academy Online, Inc., 2008, page 46.

159 University of Phoenix Web Site, "Just the Facts: University of Phoenix Background," accessed at http://upxnewsroom.com/facts/

160 "Best Worst in Online Degree Programs," op. cit., page 46.

161 Ibid.

162 "Fact Book 2007," University of Phoenix, page 8, downloaded from http://www.phoenix.edu.

163 Eve Tahmincioglu, "The Faculty is Remote, but Not Detached," March 9, 2008, *The New York Times*, accessed from http://www.nytimes.com.

164 *The Horizon Report*, op. cit., page 12.

165 Ibid, page 14.

166 Nandini Lakshman, "Online Education Takes Off in India," *Business Week*, February 22, 2008.

167 Allison Powell and Susan Patrick, "An International Perspective of K–12 Online Learning: A Summary of the 2006 iNACOL International E-Learning Survey," North American Council for Online Learning, page 9-10.

168 Ibid, page 10.

169 Katrina A. Meyer, "The Closing of the U.S. Open University," *Educause Quarterly*, November 2, 2006, pages 5-6.

170 Ibid, page 6.

171 Monica Campbell, "A Texas Institution Sees Online Learning as Growth Industry in Latin America," *The Chronicle of Higher Education*, Sept. 12, 2008, downloaded from http://chronicle.com

172 Fundación Cisneros web site, www.orinoco.org

173 John R. Vacca, "CU on the Net," *Internet World*, October 1995, page 81.

174 Davies and B. Samways, eds., "KIDLINK, Creating the Global Village" *Teleteaching*, (North-Holland: Elsevier Science Publishers, B.V., 1993).

175 From web site www.montgomerycountyschoolsmd.org.

176 Allison Powell and Susan Patrick, "An International Perspective of K–12 Online Learning: A Summary of the 2006 iNACOL International E-Learning Survey," North American Council for Online Learning, pages 8-9.

177 2008 Comscore Press Release, accessed from http://www.comscore.com/press/pr.asp.

178 Emilio Gonzalez, "Connecting the Nation: Classrooms, Libraries and Health Care Organizations in the Knowledge Age," (Washington, DC: U.S. Department of Commerce, 1995), page 4

179 "The Radicati Group, Inc. Releases Q2 2008 Market Numbers Update," Press Release on August 4, 2008, accessed from *www.radicati.com.*

180 Ibid.

181 "PCs In-Use Reached nearly 1.2B in 2008. USA Accounts for over 22 percent of PCs In-Use," Computer Industry Almanac Inc. Press Release, January 14, 2009.

182 Jeffrey Young, "I.Q. Wars," *Forbes ASAP*, December 4 1995, page 78.

183 "PCs In-Use Reached nearly 1.2B in 2008. USA Accounts for over 22 percent of PCs In-Use," op. cit.

184 "Key Global Telecom Indicators for the World Telecommunication Service Sector," International Telecommunications Union, accessed from http://www.itu.int.

185 Emilio Gonzalez, op. cit., page 10.

186 *Digest of Education Statistics 2008*, op. cit., page 611.

187 David A. Kaplan and Adam Rogers, "The Silicon Classroom," *Newsweek*, April 22, 1996, page 60.

188 Human Development Report 1999, United Nations Development Programme, (New York: Oxford University Press, 1999).

189 Ibid, page 65.

190 Ibid, page 62.

191 Dale Carnevale, "Professors take distance learning to next-generation networks," *The Chronicle of Higher Education*, 11 November 1999, A60.

192 "Frequently Asked Questions about Internet2," October, 2006, from the Internet2 main web site, accessed from www.internet2.edu.

193 "Teaching and Learning" Fact Sheet from Internet2's web site at *http://www. internet2.edu.*

194 Barry Willis, ed., *Distance Education: Strategies and Tools*, (Englewood Cliffs, N.J. Educational Technology Publications, Inc., 1994) P. 32

195 Ibid, page 226.

196 Chris Dede, "Emerging Technologies and Learning," *American Journal of Distributed Distance Education*, 1996, 10(2), pages 4–36.

197 Ibid.

198 Katrin Verclas runs MobileActive.org, a website and community of about 3.000 activest and NGO's around the world. This reference appears in "A World of Witnesses," in a "Special Report on Mobility," in *The Economist*, April 12, 2008, page 11.

199 *Measuring the Information Society: The ICT Development Index*, International Telecommunications Union, 2009, page 3

200 Ibid.

201 Bruno Giussani, "More Cell Phones Than People," LunchoverIP, June 16, 2006, accessed from www.lunchoverip.com.

202 ITU World Telecommunications/ICT Indicators Database downloaded from International Telecommunication Union web site: *http://www.itu.int*

203 "A Review of the Open Educational Resources (OER) Movement: Achievements, Challenges and New Opportunities," Report to The William and Flora Hewlett Foundation, by Daniel E. Atkins, John Seely Brown and Allen L. Hammond, February 2007, page 79.

204 *Grown Up Digital*, op. cit., page 48.

205 "A Special Report on Mobility," *The Economist*, April 12, 2008, page 3.

206 "Marketers Have Eyes on the 'Third Screen,'" by Eric Pfanner, *The New York Times*, March 22, 2007, downloaded from http://www.nytimes.com

207 "A Review of the Open Educational Resources (OER) Movement: Achievements, Challenges, and New Opportunities," Report to the William and Flora Hewlett Foundation, February 2007, by Daniel E. Atkins, John Seely Brown and Allen L. Hammond, page 75.

208 "The Next 4 Billion: Market Size and Business Strategy at the Base of the Pyramid," World Resources Institute and International Finance Corporation, pages 3, 50.

209 Ibid, page 50.

210 Janna Quitney Anderson and Lee Rainie, "The Future of the Internet III," Pew Internet & American Life Project, December 14, 2008, page 5.

211 Ibid, page 27.

212 Ibid, page 29.

213 *The Horizon Report*, by the New Media Consortium and the EDUCAUSE Learning Initiative, 2008, page 18.

214 "From Text to Context," by John Traxler, in the "Proceedings from the mLearn2008 Conference: The Bridge from Text to Context," October 7-10, 2008, Ionbridge Gorge, Shropshire, UK, University of Wolverhampton, accessed from www.mlearn2008.org.

215 YouTube, Marc Prensky based on comments made at Handheld Learning 2008.

216 Kristofor Swanson, "Case Study: Merrill Lynch: Bullish on Mobile Learning," April 2008, *Chief Learning Officer*, downloaded from *www.clomedia.com*.

217 Newswire Press Release, "Intuition and Merrill Lynch Recognized for Mobile Learning Innovation," 5/29/08, downloaded from http://www.newswiretoday.com/news/34917/.

218 "Staying the Course: Online Education in the U.S., 2008," by I. Elaine Allen and Jeff Seaman, Babson Survey Research Group and The Sloan Consortium, November, 2008, page 1.

219 "Be True to Your Cyberschool," by Alison Damast, Business Week Online, 4/20/07.

220 "For-Profit Boom Continues, Despite Enrollment Shifts," *Recruitment & Retention in Higher Education*, January 2006, Vol. 20, Issue 1.

221 "Eduventures Study Helps Colleges and Universities Understand Growing Complexity of Online Higher Education Market," Press Release by Eduventures, November 26, 2007.

222 *The Digest of Education Statistics 2008*, op. cit., page 16.

223 "OECD Thematic Review of Tertiary Education: Background Report for the P.R. of China," February 2007, page 18.

224 "Perspectives on Distance Education: Lifelong Learning & Distance Higher Education," Commonwealth of Learning/UNESCO Publishing, 2005, page 66.

225 "Perspectives on Distance Education: Lifelong Learning & Distance Higher Education," Commonwealth of Learning/UNESCO Publishing, 2005, page 65.

226 Allison Powell and Susan Patrick, "An International Perspective of K–12 Online Learning: A Summary of the 2006 INACOL International E-Learning Survey," International Association for K–12 Online Learning, November, 2006, page 9.

227 Ibid, page 10.

228 "Perspectives on Distance Education: Lifelong Learning & Distance Higher Education," Commonwealth of Learning/UNESCO Publishing, 2005, page 81.

229 Weiyuan Zhang, Jian Niu, Guozhen Jiang, "Web-Based Education at Conventional Universities in China: A Case Study," January 2002, *International Review of Research in Open and Distance Learning*, page 5.

230 *University of South Africa Annual Report*, 2007, page 3.

231 Jennifer Robison, "MBA Stands for Mexico's Business Accelerator: Why Puebla's University for Economic Development is Getting High Marks," *Gallup Management Journal Online*, 8/14/08, pages 1-7.

232 Stella C. S. Porto and Zane L. Berge, "Distance Education and Corporate Training in Brazil: Regulations and Interrelationships," *International Review of Research in Open and Distance Learning*, June, 2008, page 4.

233 Ibid.

234 Evan Ramstad, Associated Press, "Product Merges PC, TV," Boulder, Colo., March 24, 1996, page 1B.

235 Ibid.

236 Author Unknown, "Corporate Universities in the Year 2000," *Corporate University Xchange*, January-February 1996, 3.

237 "Explosive growth of broadband service over cable TV predicted," *India eNews*, November 21, 2008, from www.indiaenews.com.

238 John A. Byrne, "Virtual B-Schools," *Business Week*, October 23, 1995, page 65.

239 Cushing Anderson, "Worldwide and U.S. Corporate eLearning 2008–2012 Forecast," June 2008, Document at a Glance from IDC's web site at www.idc.com.

240 Jeanne C. Meister, *Corporate Quality Universities: Lessons in Building a World-Class Work Force*, (Butt Ridge, Ill., and New York: Irwin Professional Publishing, 1994), page 12.

241 Information from the NTU web site at http://www.waldenu.edu

242 Jones International University is accredited by the *Higher Learning Commission* and a member of the North Central Association. The Higher Learning Commission is part of the North Central Association of Colleges and Schools. The Association is a not-for-profit, voluntary, membership organization that was founded in 1895 for educational institutions. The Association is one of six regional institutional accrediting associations in the U.S.. Through its Commissions, it accredits, and thereby grants membership to, educational institutions in the 19-state North Central region. *www.ncahigherlearningcommission.org*

243 Will and Ariel Durant, *The Lessons of History*, (New York: Simon & Schuster, 1968), page 79.

244 Charlene A. Dykman and Charles K. Davis, "Part One — The Shift Toward Online Education," in the *Journal of Information Systems Education*, Spring 2008.

245 General Electric Company Press Release, "Training and Development: Our 2007-2008 Commitment" accessed from http://www.ge.com/citizenship/performance_areas/employees_train.jsp

246 William Martin, "Procter & Gamble: Break Up the Content with Surgical Learning," *Chief Learning Officer*, July 2008, pages 50-51.

247 Ibid, page 51.

248 In November, 1987, Mind Extension University (ME/U) was launched as a basic cable television channel designed to deliver advance placement (AP) courses to middle schools and high schools. It concentrated on math, sciences and languages. Its original footprint was from Alaska to Venezuela and from Hawaii to Bermuda. Later it focused on college level courses, coordinating over 30 universities so that their courses could be aggregated or integrated into degree programs. It changed its name to Knowledge TV and eventually had approximately 25 million North American subscribers in Canada, the U.S. Mexico, Central America and South America. Using satellite, terrestrial, microwave, videotape, cable TV and broadcast TV, and translating to different languages, it added 600 thousand subscribers in Poland, 200 thousand in Romania, 500 thousand in the Netherlands, small Scandinavian carriage, minor carriage in Thailand and 20 million part-time subscribers in Mainland China.

249 I. Elaine Allen and Jeff Seaman, "Online Nation: Five Years of Growth in Online Learning," October 2007, Sloan-C, page 1.

250 Ibid.

251 Della Bradshaw, "Wide range of MBA programmes offered online," March 17, 2008, Financial Times Limited accessed from www.ft.com.

252 Peter J. Derr, "Media Review," *American Journal of Distance Education*, 9, No. 3, 1995, page 85. The course is accessible at http://www.learner.org/resources/series86.html.

253 Arthur W. Chickering and Zelda F. Gamson, "Seven Principles for Good Practice in Undergraduate Education," in *The American Association for Higher Education Bulletin*, March 1987, accessed from http://honolulu.hawaii.edu.

254 Eugene Sullivan and T. Rocco, *Guiding Principles for Distance Learning in a Learning Society* (May, 1996), page 4.

255 Ibid, page 2.

256 Text of Speech by Secretary of Education Arne Duncan at the National Science Teachers Association Conference, March 20, 2009, accessed from www.ed.gov.

257 *Respectfully Quoted*, The Library of Congress, 1989, page 343 (from the "Battle-Field," by William Cullen Bryant).

258 "Teaching, Learning and Technology," a report on 10 years of Apple Classrooms of Tomorrow Research, Apple Computer, Inc., October 2, 1995.

259 Information from Apple's web site, http://newali.apple.com/acot2/principles

260 Information from University of Phoenix web site: http://education.phoenix.edu/community/alumni_profiles.aspx

261 Ibid.

262 Information from Colorado Tech Online web site, http://www.coloradotech.edu/student-life/success-stories/brackeen_s.aspx

263 Testimonial provided by Joshua Basara at Capella University.

264 David Paydarfar and William J. Schwartz, "An Algorithm for Discovery", in *Science Magazine*, April 6, 2001, page 13, accessed from www.sciencemag.org

265 *The Economist*, "Secrets of Success," September 10, 2005, page 3.

266 *The Digest of Education Statistics*, 2008, op. cit., page 271.

267 Thomas Jefferson letter to Samuel Kercheval, 12 July 1816.

268 Peter Drucker, *Innovation and Entrepreneurship: Principles and Practice*, (New York: Harper & Row, 1985), page 17.

269 Justin Lahart, Patrick Barta and Andrew Batson, "New Limits to Growth Revive Malthusian Fears," *The Wall Street Journal*, March 24, 2008, page 1.

270 Ibid.

271 Ibid.

272 Robert Thomas Malthus, "An Essay on the Principle of Population," page 4, 1798, accessed from www.econlib.org.

273 Lewis Carroll, *Alice in Wonderland*, Chapter 9, accessed from http://ebooks.adelaide.edu.au.

Breinigsville, PA USA
12 March 2010
234093BV00004B/1/P